D0251073

THE GREAT AMERICAN JET PACK

THE QUEST FOR THE ULTIMATE INDIVIDUAL LIFT DEVICE

STEVE LEHTO

First edition
Published by Chicago Review Press, Incorporated
814 North Franklin Street
Chicago, Illinois 60610
ISBN 978-1-61374-430-7

Library of Congress Cataloging-in-Publication Data

Lehto, Steve.

The great American jetpack : the quest for the ultimate individual lift device / Steve Lehto. — First edition.

pages cm

Includes bibliographical references and index.

ISBN 978-1-61374-430-7 (hardback)

1. Personal propulsion units—United States—History. 2. Aeronautical engineers—United States—Biography. I. Title.

TL717.5.L44 2012

629.1'4—dc23

2012039337

Cover design: John Yates at Stealworks.com
Interior design: Jonathan Hahn

Printed in the United States of America
5 4 3 2 1

For
Milo and Wolfy,
Brandy and Sangria.

CONTENTS

INTRODUCTION

In his 1887 book *The Clipper of the Clouds*, Jules Verne described a fictional meeting of scientific gentlemen debating whether man might fly, and if so, how. A stranger appeared before them and announced that lighter-than-air contraptions such as hot air balloons were impractical. The bigger they were, the harder they were to maneuver. No, the answer lay in heavier-than-air flight. Over the clamorous objections of his learned audience, the man boldly predicted:

> Yes, the future is for the flying-machine. The air affords a solid fulcrum. If you will give a column of air an ascensional movement of forty-five metres a second, a man can support himself on the top of it if the soles of his boots have a superficies of only the eighth of a square metre. And if the speed be increased to ninety metres, he can walk on it with naked feet. Or if, by means of a screw, you drive a mass of air at this speed, you get the same result.[1]

A man walking in the sky? In August 1928, the science fiction magazine *Amazing Stories* featured a cover with a colorful drawing of the Skylark of Space, a futuristic character hovering a few feet off the ground, held in the air by a small device on his back.[2] The image was the product of the artist Frank R. Paul, who often illustrated stories of flying saucers from outer space and battles that raged on the surfaces of other planets.[3]

The Skylark of Space has been largely forgotten but inventors did build a working model of the device on his back. It would eventually creep into the consciousness of America, and people across the United States saw men flying in the same manner as the Skylark of Space, zooming around effortlessly in the sky as if gravity had finally been made obsolete. The men, wearing what most people would mistakenly call jet packs, became ubiquitous in the 1960s and 1970s. The devices were featured in movies and television shows and hovered over Super Bowl halftimes and Olympic opening ceremonies. They became known worldwide. But then they seemingly vanished from the landscape. Only a few isolated stories would pop up from time to time of inventors reviving the technology, or diehard enthusiasts working to keep the technology alive, flying the devices in demonstrations and at public events.

The story of the jet pack is really the story of man's dream of flying. Not in the manner of the Wright Brothers, who built a huge flying machine that could take off and fly with more than a little bit of effort. The jet pack—or the individual lift devices, as they were blandly labeled by the government men who financed much of their development—answered man's desire to simply step outside and take flight. No runways, no wings, no pilot's license required. Soaring through the air with the wind in your face and landing anywhere there was room to stand. Could it be done? Yes, it could be done, and it was.

A vehicle that takes off vertically, like a helicopter, is described as vertical take-off and landing, or VTOL. The personal flight devices covered in this work are, for the most part, vertical take-off and landing devices. This attribute separates them from small airplanes and gliders, and makes them closer to the notion that one could simply step outside and take flight. VTOL vehicles do not require landing strips or airports. They just need a place to land and a view of the sky.

To the average person, the term *jet pack* is often used to describe any individual lift device, regardless of its means of propulsion. To a purist, the first such device that flew safely and practically was called a rocket belt. It was powered by a chemical reaction that created thrust and it flew for less than half a minute. Later models powered by small jet engines flew longer and were real jet packs, but they were deemed jet belts to keep in line with the belt terminology.

Although early developments in the field showed promise, the technology stalled. The practical jet pack seemed tantalizingly close but always just out of reach. When one problem was solved, another would replace it. The only constant was the promise—the promise that jet packs would soon be here, available for everyone. But soon never came.

The story of individual lift technology truly started with the advent of flying platforms in the 1940s. Shortly after the platforms came the rocket belts, and then the jet belts. Innovation in these designs followed an interesting trajectory. The individual lift devices morphed from carrying a passenger to being worn by the passenger. Later they came full circle, returning to something a passenger stood on to be lifted skyward. Innovations in the twenty-first century even allowed a man to fly across the English Channel wearing a descendant of the jet belt.

A common thread among these devices is that the pilot and the machine were usually connected, often attached, to each other. Further, these flying machines were controlled to a large extent by the movements of the pilot, something called *kinesthetic control.* While airplanes and helicopters required trained pilots to work levers, pedals, joysticks, and yokes, the individual lift devices often had nothing more than a handle or two—the pilot simply leaned or twisted in the direction he wanted to travel. It was this last point—the ease with which an untrained pilot could take to flight—that was often the major selling point made by the promoters of this technology.*

This promise of ease in operation resulted in eventual disappointment for the people who hoped to fly like this someday. The jet pack was perhaps the most overpromised technology of all time. Creators of the flying platforms, rocket and jet belts, and the later flying devices and the men who flew them audaciously promised that one day, we would all have them.[4] They would be available for everyone, affordable, and they would be as commonplace as automobiles. They would not only fulfill humankind's dream of flying, they would allow us a magnitude of freedom hith-

*This is an admittedly artificial distinction being made by the author. There were also one-person helicopters and hovercraft that could be considered similar to the devices being written about here. However, for the purposes of this work, the author is sticking to what appears to be a logical class of devices: those that appeared simplest to operate and offered the average man the ability to fly "without wings."

erto unknown to modern man or woman. No more traffic jams; no need to rely on mass transit. Soon, it was promised, you would be able to simply strap on your jet pack or step onto your own flying platform and zoom to wherever you wanted to go.*

And then, in the twenty-first century, the largest strides were made in the development of these devices. While engineers and hobbyists revisited the technology of the rocket belts, doing what they could to tweak more than a half minute of flying time from units modeled on the 1950s versions of the rocket belt, a Swiss pilot and inventor named Yves Rossy built a flying apparatus that came perhaps the closest yet to man's dream of free flight. His creation was a small wing fitted with miniature turbine engines, technology that did not exist in the 1950s. "Jetman" Rossy crossed the English Channel and flew over the Alps with his jet pack using nothing more than his body movements to steer. Perhaps the dream of individual flight is within human reach after all.

---

*The term jet pack has become synonymous with an unkept promise of a better future through science. There is even a book entitled *Where's My Jetpack? A Guide to the Amazing Science Fiction Future That Never Arrived*, by Daniel H. Wilson (2007). That question has been asked by many. Another writer from the *New York Times* asked "Where Are Our Jetpacks?" in "Canceled Flight," *New York Times Magazine*, June 11, 2000.

1

FLYING SHOES AND HOVERING PLATFORMS

In the fall of 1942, a thirty-four-year-old engineer named Charles Horton Zimmerman was thinking about flight and what it would take to make it available to everyone. He had been working at Chance-Vought, a manufacturer of military aircraft, as well as at the laboratory of the National Advisory Committee for Aeronautics. For an aeronautical engineer, he held an unorthodox belief: he thought airplanes were too difficult to fly. He knew that trained pilots could fly them, but he hoped aviation could be pushed to a point where anyone could fly. "I couldn't see any future for personal aviation. There seemed to be no prospect of a plane *I'd* want to fly." Zimmerman wondered how to simplify the flying machine. What was necessary to lift a man into the air? Reduced to its components, it wasn't much. An upward force pressed against his feet would lift a man if the force was greater than that of gravity. Could an individual, simply being lifted, remain stable in "flight" like this? If so, it was simple. There it was: lift and control. Two problems Zimmerman hoped to overcome to allow the average person to fly.

Zimmerman knew that stability in the air was the real problem. He had been working on airplane stability in wind tunnels for more than a decade and was an expert on the subject. Like so many inventors, he opted to work out of his garage while attacking the problems of personal flight. He hung ropes from the rafters and placed a board between them.

He put a stick under the board and then stood on it to see how hard it was to keep his balance when perched on such a small platform. He found it was much easier to keep his balance than he had expected. Could he balance on a platform if it was lifting him into the air? He thought so. He set out building a device he called his "Flying Shoes."[1]

Zimmerman thought it might be possible to get the lift he needed using miniature helicopter blades.[2] The device he built consisted of a small platform on a framework of tubing that stood just a few inches off the ground. Zimmerman, as the pilot, would stand between two small four-cylinder engines powering two-bladed propellers that faced upward. The spinning blades were about knee-high to Zimmerman and controlled by a throttle handle extending upward from the base.

As he worked on developing the Flying Shoes Zimmerman became convinced that he could not only balance on the device when it was aloft, but that he could also maneuver it with kinesthetic control.[3] That is, he believed he could stand on the Flying Shoes and simply lean in the direction he wanted to go. If he leaned forward, it would tip the device and the change in the angle of the propellers would move the platform forward. To stop, he thought he could just lean backward. If true, this meant that an operator would not need special training to pilot the Flying Shoes. The flight controls were intuitive. This would be the flying machine he had dreamed of, the one that could be flown by anyone.

Zimmerman filed for a patent on the Flying Shoes in 1943. The patent was granted in 1947, and it is clear Zimmerman had some great aspirations for his invention. Patent filings often contain quite broad and general descriptions of an invention and sometimes appear primitive when compared to the products that follow. Zimmerman's patent for a Helicopter Flying Apparatus covered the Flying Shoes, but it was outlandishly different from what he had built. In his patent, Zimmerman imagined the flyer attaching himself to the machine, which had blades surrounded by rings. The pilot stood directly on top of the propeller housings, which—when compared with the open-bladed Flying Shoes prototype—seemed a much safer design. The drawings in the patent show that Zimmerman did not simply intend for his device to hover. He drew small wings on the back of the pilot and even smaller wings on the side of the operator's head.

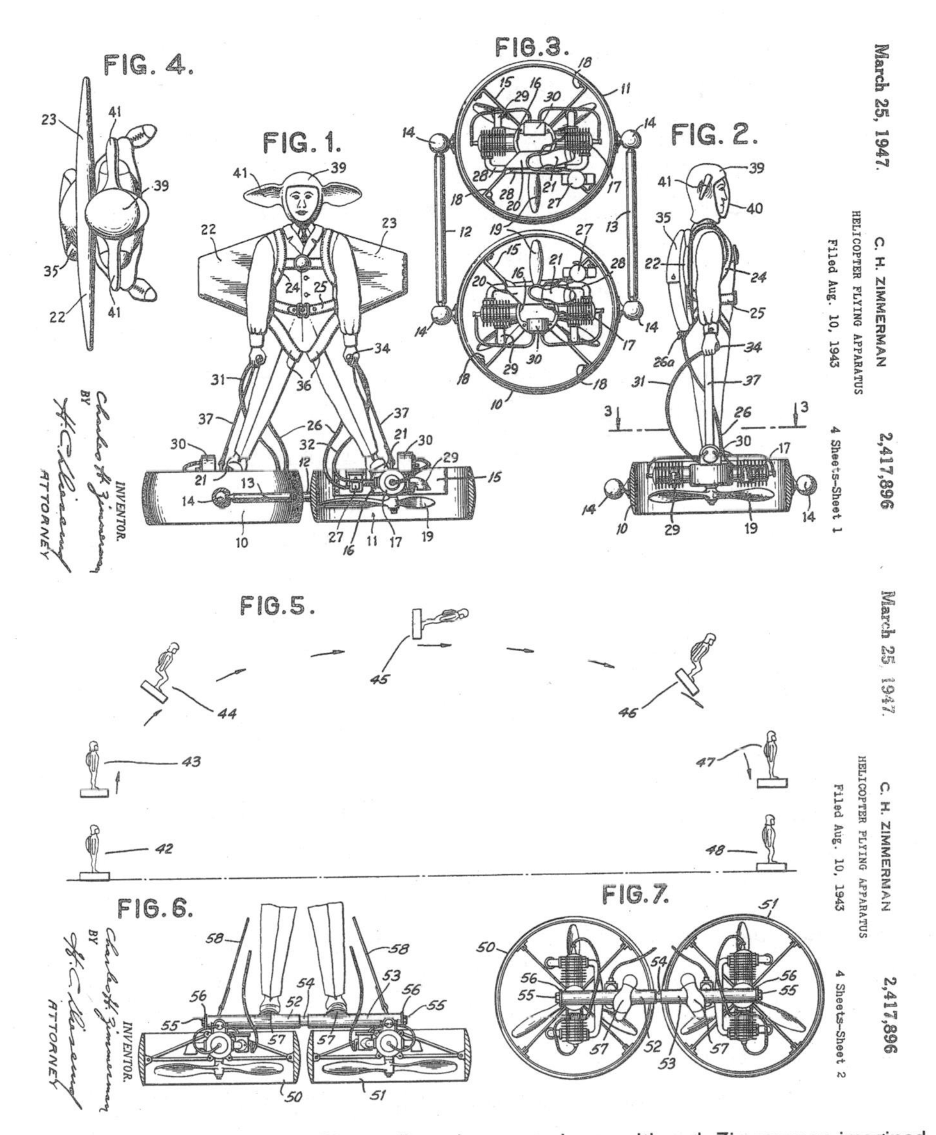

Charles Zimmerman's Flying Shoes allowed a man to hover, although Zimmerman imagined that the pilot would be able to fly horizontally as well. While the machine never fulfilled that dream, it did demonstrate kinesthetic control.

An accompanying illustration showed a typical flight: the flyer would lift off the ground, tip forward, and then fly in the manner of an airplane, using the wings on the flyer's back and head. When he wanted to land, he would slow down and swing his legs beneath him and then land vertically. Zimmerman had never gotten his shoes to do any such thing, and he never would. The claims he made in his patent for the device were likewise impossible. He said the device "should be able" to travel 350 miles at speeds approaching two hundred miles per hour.

Notwithstanding the outlandish performance claims, Zimmerman's patent contained real elements that made it groundbreaking. He proposed a device to be worn by its pilot, with vertical take-off and landing, that would be controlled kinesthetically. The only mechanical control for the pilot was the throttle. Zimmermann went so far as to suggest his device could be powered someday by a turbine engine.[4]

Even though Zimmerman worked in an aeronautical laboratory, he chose to build the Flying Shoes in his garage using his own money. The engines cost him $500 each and the propellers were $100 for the pair. Zimmerman, like many other inventors, insisted on testing the device himself. A magazine reporter remarked on Zimmerman's bravery, standing on the Flying Shoes with the wooden propellers whipping by just four inches away from his knees.

While Zimmerman was fine-tuning his Flying Shoes, he met Stanley Hiller Jr., a twenty-one-year-old businessman and inventor who built and flew his own helicopters and had, at the age of seventeen, established Hiller Industries in Palo Alto, California. Hiller's company, later known as Hiller Aircraft and then Hiller Helicopters, developed helicopters for the military and civilian markets.[5] Hiller was always looking for new developments and inventions. While he was traveling on the East Coast in 1946, Hiller heard about an inventor working on a "crazy machine" nearby.[6] The man telling Hiller the story thought Hiller would get a kick out of it since the machine contained elements of a helicopter. Hiller went and visited Zimmerman to see what he had created. He was intrigued.

Hiller brought the Flying Shoes to California for testing.[7] It didn't go well. It was hard to keep the two engines running at the same speed and, as a result, the device proved unstable. Plus, what would happen if one of the engines failed in flight?[8] Worse, how safe was it for the operator to be standing so close to two propellers spinning at full speed? And the machine never got far off the ground; it hovered at an altitude of a few inches.[9] Hiller sent the Flying Shoes back to Zimmerman but remained intrigued by the idea.

Meanwhile, Zimmerman remained fixated on the idea of kinesthetic control, convinced that a person could balance and control a hovering device using nothing more than body movement. But how could this be proven? The National Advisory Committee for Aeronautics agreed to let Zimmerman work on the problem using its facilities. The scientists and engineers there routinely used large quantities of compressed air stored in giant tanks. Zimmerman thought he might be able to build a rig where a small platform would be blasted upward using nothing more than the compressed air. On February 2, 1951, under his direction, Zimmerman's engineers attached air hoses to a piece of plywood with a small nozzle in its center. The hoses fed air through the nozzle at several hundred pounds per square inch, and Zimmerman's feet were strapped to the board. They connected safety tethers to him, in case it didn't work. Then, they let the air loose.

The air shrieked through the nozzle and Zimmerman waited to see what would happen. Nothing. Then he looked around and realized that his safety lines were no longer taut. He later said he was hovering off the floor but hadn't felt the board move. He stood motionless for a minute and then signaled for the air supply to be cut off. He had proven he could remain stable on a column of air, but could he maneuver? Zimmerman and another man named Paul Hill took turns flying the board and studying the limits of hovering flight. They realized that Zimmerman had been correct: they could move about on the column of air by simply leaning in the direction they wanted to go. "Paul Hill became adept at sashaying around the 15-foot circle to which the dragging lines limited him." Soon others were riding the board on a cushion of compressed air.[10]

Paul Hill liked the idea so much he built a contraption that looked like a double-bladed, upside-down helicopter. It weighed 140 pounds, but it flew. The blades were powered by compressed air. Its seven-foot propellers were on the same axis but spun in opposite directions. It responded to the pilot's body movements, even if it wasn't as responsive as the flying board. Word of these NACA experiments found their way to Hiller, who had not forgotten the Flying Shoes.[11] Hiller was so convinced the idea was viable that he began looking for a government sponsor.

2

THE HILLER AND DE LACKNER FLYING PLATFORMS

In 1953, the Office of Naval Research–Naval Sciences Division heard about the NACA work and Stanley Hiller's desire to create a practical individual lift device. It offered Hiller a contract to develop a ducted-fan flying platform, suggested by a government scientist named Alexander Satin who had followed the work of Charles Zimmerman and Paul Hill.[1] Zimmerman reportedly waived any patent rights he might have in whatever Hiller developed.[2] The proposed device consisted of two large propellers that spun, one in each direction, inside a round fiberglass housing that was a foot or so tall. On top of the housing was a small spot for the pilot to stand, surrounded by a frame of aluminum tubing to keep the pilot from falling off the device. In essence, the device was a huge fan pointed at the ground. The pilot would stand on top of the platform and when it lifted itself, it would lift the pilot with it. The flying platform would be controlled kinesthetically.[3] In many respects, it was a larger version of the Flying Shoes, slightly reconfigured. An advantage of this new design was that the duct—the circular shroud that surrounded the propellers—concentrated the air more efficiently than a non-shrouded propeller. Hill and Zimmerman consulted with Hiller on the design, and the project was developed and built under strict secrecy.[4] While Hiller Helicopters employed nine hundred people, only fifteen knew about the flying platform before it was unveiled publicly, according to Hiller.[5] The work began in January 1954 and Hiller had a working prototype before the end of the year.[6] The device was five feet across and two men could lift and carry it. It became the "first ducted fan VTO vehicle to fly successfully." It has also been claimed that it was the "first heavier-than-air aircraft capable of being flown from the outset by someone without flight training," although it is unclear if anyone who was not a pilot ever actually flew the Hiller Flying Platform.* While the flights were not spectacular, they highlighted the simplicity of kinesthetic control. The VZ-1 Flying Platform made its first free flight January 27, 1955.[7]

By April 1955, Hiller was ready to let the world know about its new flying machine. Hiller reported it to *Flight* magazine, an aerospace periodical that acted as a sort of yearbook for the aviation industry. Although the magazine covered all manner of aircraft, the editors found the flying platform unusual. "Retrospectively, the past twelve months seem to have introduced—albeit in rudimentary forms—an unprecedented number of curious types of flying machine. One of the strangest of all is [the Hiller Flying Platform]."[8] *Flight* cautioned readers not to expect to see the flying platforms in widespread use just yet; the flying platform was a "research tool," and "further research and development will be necessary."[9]

Hiller publicized the new invention in the mainstream press also, and as would become commonplace, the new individual lift device was introduced to the general public with overblown superlatives, setting it up to almost certainly disappoint. It could never live up to the hype. Reporters from *Collier's* were given a demonstration of the VZ-1 and could barely contain themselves, gushing about the platform and the impact it was sure to have on society. "A radically different one-man aircraft, it hovers, climbs and darts sideways, 'riding a column of air.' It's probably the simplest flying machine ever created—and it may revolutionize aeronautics."[10]

Hiller touted the machine as being equally valuable to the civilian market as it might be to the military. "It is not inconceivable that it might become the long-awaited 'airplane in every garage,' which never became a reality because of the high cost of planes and helicopters. Simplicity is

---

*Keep in mind that there is a distinction made by using the term "aircraft." The rocket belts were powered by rockets and were not considered aircraft. Later, jet belts would be considered aircraft.

the keynote of the Hiller-ONR machine, and simplicity means lower cost. Anyone with a sense of balance can learn to fly such a vehicle. If you've ever ridden a bicycle you're a platform pilot prospect." A project engineer gave the reporter an even better standard: "A trained bear could fly this machine."[11] Further, it would work wonders for the American fighting man. It could be used to assault beaches, cross rivers, or patrol "otherwise inaccessible areas."[12]

Stanley Hiller Jr., the president of Hiller Helicopters, was asked to explain to the readers of *Collier's* how kinesthetic control worked. He did his best while overselling the idea in the process. "In essence the pilot is standing on, and riding, a column of air. Stepping into this machine is like stepping into your shoes. As you walk with your shoes, you can fly with this platform. Since the early gliders, prior to the flight of the Wright Brothers, there has never been a flying device which depended on the natural reflexes of the body for its control. Now we've reached instinctive flight. We're *putting on* the machine, instead of getting into it. In the future, if you can stand, you can fly."[13] Although he was exaggerating how easily the flying platform could be flown, he was onto something: kinesthetic control was revolutionary and it was possible.

Although Hiller claimed that the flying platform did not require pilot training to fly, Hiller only demonstrated it being flown by their test pilot, Phil Johnston, who was a World War II fighter pilot before coming to Hiller Helicopters. He said that the device was simple to fly and easy to master and that it only took fifteen minutes of training. It was so easy to fly, "a child could control it."[14] Hiller Helicopters never mentioned whether they had ever actually let someone untrained try to fly the device. If it was as simple as putting on one's shoes, why hadn't they tried it?

Buried deep within the article were details that revealed some of the flaws in the device. So far, the longest flight of the VZ-1 had lasted just three minutes. The engineers conceded weaknesses in the machine's design. The propellers were belt-driven. If the belts snapped or came off, the platform would fall. The device had two engines and each drove one of the propellers. If one engine stopped, the device would spin out of control due to the unbalanced torque of the remaining propeller. The fear of losing power in flight would haunt the field of individual lift. Airplanes and helicopters can, to varying degrees, land safely if they lose power from a complete engine failure. The flying platform, much like the other individual lift devices that followed it, could not safely land from any considerable height after loss of power. This problem was euphemistically called "lift degradation" by engineers; others might characterize it simply as falling out of the sky, or "crashing."[15]

The flying platform was still being refined, but Hiller didn't want any of these caveats to scare off future purchasers. He said the vehicle was "as safe as the family automobile." Reporters were told the machines would start out priced in the range of $1,000 to $2,500 each, but with mass-production techniques, they could be churned out "like washing machines," at only $500 apiece.[16]

Hiller mentioned one more detail that was probably the real reason he decided to unveil the previously top secret project to the press in the first place: these devices could not become available to the public without continued financial support from the military. The platform was still experimental, but with further funding Hiller hoped to refine it and make it available to consumers. Hiller also told the reporters about other possible evolutions of the ducted fan method of propulsion. He could build bigger devices and even aircraft that could take off like helicopters but fly like airplanes. It was all a matter of funding. Still, "anything that man can think of, man can do."[17]

Flight magazine published an updated synopsis of the Hiller Flying Platform program in its November 1956 issue. They, too, were impressed by the concept of kinesthetic control. The platform "is almost literally a flying carpet which the pilot controls by body balance instead of using manual or mechanical flight controls. Directional flight is achieved merely by leaning in the direction one wishes to go."[18]

Although the Office of Naval Research had funded the initial development of the flying platform, they decided not to order any more of the devices from Hiller. However, the army liked the idea of the VZ-1 enough to order three flying platforms in 1956. They asked for the platforms to be modified a bit. Worried about the dangers associated with losing an engine in flight, they asked Hiller to add a third engine. The additional engine added power but also increased the weight of the platform. To lift the heavier components, the platform needed larger propellers. The new VZ-1E was eight feet across and sported three forty-four-horsepower four-cylinder engines.

When Hiller told *Flight* about the improved model, the company admitted that the larger platform was harder to steer. The pilot could no longer count on the machine simply going in the direction the operator was leaning. Hiller installed four adjustable vanes underneath the propellers to help steer the craft, but it didn't mention how the operator got them to adjust. Presumably the additional steering devices needed their own controls and wouldn't respond to the pilot's shifting weight.[19] Hiller didn't release much more information to the magazine about this aircraft, but instead pointed to devices Hiller said it was developing, such as "twin fan platforms and flying cranes."[20]

Hiller built three of the VZ-1Es and delivered two to the US military. The other one the company kept.[21] Hiller wrote a classified report for the army to explain some of the issues that arose with the flying platform and how they might be solved: "Stability analyses of flying platform in hovering and forward flight."[22] Attempting to smooth out the flight characteristics of the craft, engineers had tried putting springs under the pilot's feet. It didn't help.[23] They installed a gyrobar that sensed the movements of the platform and corrected them by moving the vanes under the platform to compensate. This addition helped with control when the platform was hovering in one place. Once the platform started flying in any direction, the gyrobar didn't help as much. Hiller suggested further study, presumably at government expense.[24]

Hiller's testing with the gyrobar had been conducted on the smaller flying platform. The new larger flying platform did not perform as well as the army expected and the flying platform program was quietly ended. Even so, Hiller filed for a patent on the flying platform on February 27, 1956. The Vertical Take-Off Flying Platform application described the earlier of the flying platform configurations as a "wingless aircraft" with two engines but suggested that it could just as easily be configured for more engines or just one.

Calling it a wingless craft was a misnomer. The duct surrounding the two propellers was actually a functional wing; it just wasn't a typical airplane wing. The top of the duct had a lip that curved outward; when viewed in a cross section, it formed an airfoil. As air was pulled through the duct and drawn downward, much of it first passed over the curve at the top of the duct. This air passing over the edge created lift just the same as an airplane wing and contributed as much as 40 percent of the flying platform's lift, according to Hiller.[25]

Perhaps the most interesting feature of the patent in regard to individual lift devices was that it described kinesthetic control as a primary feature of the platform. Hiller spelled it out early in the patent's claims. The flying platform flew "under such conditions that directional control, and transition from vertical to horizontal flight and vice versa are affected by body movements or balance of the pilot flying the machine." The patent, #2,953,321, was granted in September 1960.[26]

Hiller was not the only manufacturer working on flying platforms around this time. Another company named de Lackner also built a small platform that lifted its pilot with spinning blades. Its inventor was Lewis McCarty Jr., an engineer who had also heard about the NACA tests. He built a prototype the same year Hiller built its first flying platform and flew his successfully in January 1955.[27] The blades in the DH-4 Aerocycle were powered by a forty-three horsepower Mercury outboard boat motor adapted for the purpose, but no duct or shroud surrounded them as in the Hiller devices. The blades underneath the pilot were fifteen feet from tip to tip.[28] This design gave the impression that the pilot was in constant peril of being diced by the propellers if he were to stumble in flight. Even so, de Lackner claimed speeds of sixty-five miles per hour and photographs showed the machines hovering more than twenty feet off the ground. Like the Hiller platforms, the de Lackners were steered kinesthetically.[29]

The army was impressed and ordered a dozen for further tests. In 1956, a captain named Selmer Sundby tested the Aerocycle for the military at Fort Eustis. Sundby was an experienced helicopter pilot with over fifteen hundred hours of flight experience. Although he managed to keep it aloft for forty-three minutes on one occasion, he said, "It only took me one flight to realize that a non-flyer would have considerable difficulty operating it."

At least one other person flew the Aerocycle. Publicity photos were provided to the media entitled, "Army tests one-man helicopter." According to the caption, the soldier flying the device in the photo was Sergeant First Class Herman Stern, and the vehicle was being tested at Camp Kilmer, New Jersey. In the photo of the airborne vehicle, Stern was in full military gear, with a rifle and rucksack on his back, a canteen on his belt, and a GI helmet on his head. His perch, a small step not much wider than his boots, looked precarious with the two giant propellers spinning below him. The accompanying text explained, "The pilot guides the device by leaning in the direction in which he wishes to travel." The makers claimed the Aerocycle could travel 150 miles without refueling.[30]

Sundby had a couple of close calls in the Aerocycle; one time he crashed from a height of forty feet. Another time he crashed while the unit was tethered. Both crashes had the same cause, and it wasn't operator error. The Aerocycle's two blades spun in opposite directions. The blades were very close together, and under certain conditions the blades would flex. When they flexed too far they struck one another with catastrophic results, crashing the unit and destroying the blades. The army didn't bother to figure out what the thresholds were or if there was a way to operate the Aerocycle without the blades hitting each other. It abandoned the Aerocycle program.[31]

Before the de Lackner and Hiller devices were scrapped, they were both featured in an article in *True* magazine, "The Sky-High Invention," which outlined all the developments up to 1956 with an optimistic view of the flying platform's future. "You may soon have your own fly-it-yourself air scooter—at far less cost than a family car." That overpromising line ran directly below a photograph of Zimmerman standing on his Flying Shoes in the driveway of his home. It's hard to tell from the picture but he might have been hovering an inch or two off the ground.[32]

Sundby was decorated with the Distinguished Flying Cross by the army in 1958 for his flights aboard the Aerocycle.[33] By this time though, the aviation experts and the military were turning their attention from propellers to rockets. The development of the flying platforms would pay huge dividends later, however, because of the discovery that kinesthetic control actually worked.

3

THE AGE OF MAN ROCKETS

A variety of stories are told of the origins of rocket belt technology and how it first caught the attention of the military. One involved an army officer named Colonel Charles Parkin at West Point in 1940 who was training men to use flamethrowers. At the time, flamethrowers used tanks of nitrogen to propel the flammable liquid, and one of Parkin's soldiers accidentally opened a valve on a nitrogen tank he thought was empty. He quickly closed the valve on the hissing tank and asked Parkin what would happen if the valve were opened all the way with a full tank. Parkin told him, "It would probably go skittering across the field."

The soldier jokingly asked, "Could I hang on and get a ride?"[1] Parkin wondered about it.

After the war, he remembered the exchange and decided to try a little experiment. He said he found one of the nitrogen tanks and strapped it to his back. He opened up the valve and jumped. He felt that the thrust of the tank aided his jump, more than enough to make the project worth pursuing. After all, the tank was heavy and there might be more efficient things to power such a device. What if some of our best scientists and engineers worked on this notion? By 1957, Parkin was working in the army's Transportation Research and Engineering Command, known as TRECOM, and he had not forgotten about his experiments with compressed nitrogen.

With the advances in rocketry that had been made during World War II, a lot of people had been thinking about strapping themselves to rock-

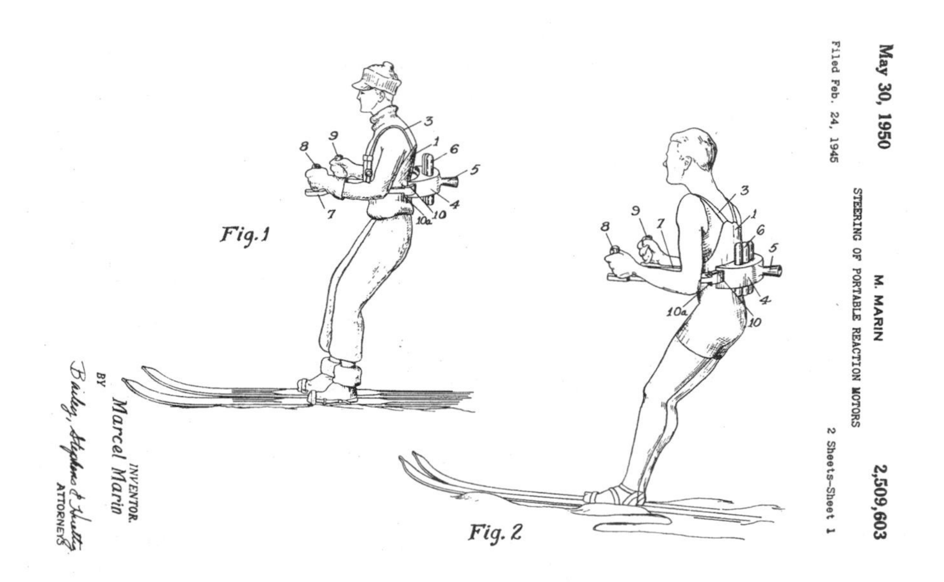

In 1945, an inventor named Marcel Marin applied for a patent on his idea of strapping a rocket onto the back of a man on skis, one of the earliest American patents for a man-rocket configuration.

ets or vice versa; it is hard to say who had the idea first. The idea very well could have been so obvious that it spontaneously occurred to several different people. In 1945, a French inventor named Marcel Marin filed a US patent for "Steering of Portable Reaction Motors." While the title may not suggest a jet pack, the accompanying illustrations on the patent do. A man is shown on skis—both on water and snow—with a small rocket apparatus attached to his back with a harness. The inventor was clear that his device would likewise propel a person on roller skates, a bicycle, or in a "light car." Clearly this was not a flying device, but the similarities are obvious enough. Marin did not specify what type of motor his invention would carry or what kind of fuel it would use but gave an array of possibilities including liquid or solid fuel rockets and compressed gas.[2]

Marin's was not the first patent for a wearable rocket. A Russian scientist had filed an interesting patent long before World War II. In 1929, Nikolai Alekseevich Rynin wrote a book titled *Interplanetary Flight and Communication*. The work was speculative in nature and contained a survey of sorts, showing all of the steps taken by man to fly, including the rudimentary flying machines, balloons, and rockets developed by this time. The book was also filled with fantastic products of the imagination, things people had thought of but not actually developed. Many of them were mythical, such as the Chinese inventor who is said to have blown himself up several thousand years ago trying to fly a rocket-powered kite. Buried deep in the book's second volume was a description of something called "Andreev's Rocket Apparatus," the subject of a patent in Russia granted to a man named A. F. Andreev. Perhaps due to the oddities of 1920s Russian patent law or due to bad translations of *Interplanetary Flight*, the patent was said to have been granted in 1928, retroactively to 1924, even though it had been filed in 1921. The invention was described as "a portable rocket apparatus which . . . a person could carry on his back like a knapsack." It was to be powered by methane and oxygen and was "intended for the transportation of a person or small loads for a distance up to 20 km." According to the patent, it would also travel at around 120 miles per hour, and only weighed one hundred pounds. The fuel only weighed sixteen pounds. It is worth noting that the proposal suggested the device could travel *twelve miles* on a single load of fuel. If nothing else, this patent proved one thing: exaggeration in this field had begun by 1921.[3]

Rynin's book included a copy of the patent illustration but it would have been little use to anyone curious about the device's planned operation. Its description said that a person could wear the device and control the "angle of inclination of the gas jets in relation to the apparatus as a whole."[4] There was no hint that Andreev ever tried to build his device, nor any explanation of how he derived his projected performance figures. Andreev's only contribution to this field seems to be that someone else saw his patent and wondered if a man could strap a rocket to his back and live to tell about it.

Thomas Moore was a scientist who worked closely with Werner Von Braun in the 1950s. Along with other rocket scientists, they were making huge strides toward putting a man on the moon. Maybe they could also build a small man rocket? Von Braun told Moore he liked the idea and in 1951 he helped Moore lobby the government for development money. The army gave Moore $25,000 to flesh out the idea. Moore came up with a design that would be powered by a mix of hydrogen peroxide, ethyl alcohol, and liquid oxygen.[5] He later told people he had been inspired by the Andreev patent.

Moore imagined a "Jet Vest" with fuel tanks that would be worn on the back. Fuel from those tanks would enter a combustion chamber and a chemical reaction would blast exhaust out through vent tubes that pointed downward, behind the pilot. The exhaust would rush out of the tubes with such force that it would lift the pilot wearing the Jet Vest off the ground. Moore built a mock-up of the vest for testing. Before he built the motors, he needed to know if the Jet Vest would fly and if it would be controllable. He built the plumbing portion of the Jet Vest and connected it to high pressure air hoses, much as Zimmerman and Hill had used compressed air to power their hover board. He then strapped himself in and blasted away while attached to safety tethers. As far as we know, he never managed untethered flight with the Jet Vest. He also never configured the vest to be operated with the rocket fuel motors. After the money ran out, he apparently didn't have enough to show the army and no further funding was given.

Did Moore see Andreev's patent in Rynin's book? If so, he may have wanted to heed a warning on the first page. Rynin noted that the science of rocketry went back thousands of years but rocket designers still encountered numerous technical problems. Rocket engines developed such high temperatures that the rockets often destroyed themselves when they flew. Rockets flew very fast but often only for short periods; the fuels were dangerous; the engines often produced too little power to be practical; and the contraptions were difficult to control.[6] Could Cold War–era scientists overcome all of these obstacles to make a wearable man rocket?

As is often the case in modern technology, engineers and scientists at different companies were working on the problem independently of each other. A company called Thiokol Chemical Corporation—Reaction Motors Division announced it had been working on a rocket device for a man to wear. The company called it a jump belt. Initially, Thiokol had made solid-fuel rocket engines and Reaction Motors had been making liquid-fuel rockets. Thiokol bought Reaction Motors in 1958, and by the end of the year the new conglomerate announced a groundbreaking invention it had teamed up to build. In presenting its device, the company claimed to have had conducted some interesting experiments. According to Thiokol, tanks of compressed nitrogen were strapped to a man's back in a configuration that allowed him to blast 350 pounds of thrust for five seconds out of small nozzles at his waist. Thiokol claimed that its test pilot had managed to jump thirty feet horizontally and fifteen feet high using this nitrogen-powered contraption. It is unclear how a man could jump fifteen feet up without getting hurt on the way down. Did the system also break his fall somehow? Thiokol only produced photographs of a man blasting a large cloud of dust at an altitude of perhaps three feet. But more interesting was the company's claim about the man running with the unit on his back: Thiokol said that he hit twenty-two miles per hour, albeit briefly, when he fired the unit for five seconds while simply running along the ground. Again, it seems hard to believe. How did the man's feet keep up with the jet blast? How did he slow down and stop afterward? Thiokol was long on promises and short on evidence for the press, but the press played along, publishing the performance numbers without questioning them.

In June 1958, Thiokol had shown their nitrogen unit to the army. The military was curious enough to ask Thiokol to build something that would run longer than five seconds. Thiokol built a unit powered with hydrogen peroxide, one of the liquids Thomas Moore had used in his experiments. This was successful, but Thiokol only considered using its jump belt for augmenting a man's jumping or running ability. The company did not imagine the invention had the ability to hold a man in the air once he had jumped.[7]

Thiokol shelved the hydrogen peroxide unit and instead developed the jump belt, powered by little rocket motors. *Popular Science* magazine reported on "Man's first leap toward free flight," and boldly stated, "Man's age-old dream of flying like a bird, free of any clumsy machinery, may be nearer than we think." The article described a belt holding five canisters that sat against a man's back. On each side of the belt were nozzles that aimed downward. The canisters contained solid rocket fuel that burned for a few seconds, and the resulting jet blast was ducted to the side nozzles. The canisters could be fired one at a time or all at once. In sequence, they would give perhaps a minute of boost. Thiokol told *Popular Science* that the exact facts and figures of the device were classified but the company made it clear that it was on the verge of a breakthrough. Thiokol told the writer that a man wearing this belt could "broad-jump a 50-foot river," "leap from the ground into a second-story window," or "race off at 35 m.p.h."[8] It is important to note that no one had ever done any of these things while wearing one of these devices.* And, once again, a manufacturer of an individual lift-type device resorted to exaggeration to try to sell an invention.

Thiokol pitched the solid-fuel rockets for use in slowing the descent of paratroopers. The company theorized that a parachuting soldier could wear one of the jump belts and fire the rockets a moment before he hit the ground. This would enable the paratrooper to soften the impact of an otherwise potentially hazardous landing. As odd as this sounds, the United States later discovered that the Soviets utilized just such a rocket braking system when dropping large payloads of cargo by parachute.[9] They did not use them for individual soldiers, however.

Thiokol told the reporter that the rest of the story was classified. It is unclear if the writer asked how far the experiments had progressed with the solid-fuel jump belt. Thiokol gave *Popular Science* photographs of a man wearing "a crude early version of the one-man jump belt" and claimed that the photographs showed the man "Breaking the world's broad jump record (26 feet 8¼ inches)." The photographs are grainy, on the order of pictures provided of alien autopsies or Bigfoot sightings. Still, simple comparison of the photo of the man jumping and the one of the man landing make it clear that, at least in these pictures, the man did not clear twenty-six feet. It appears that the *Popular Science* writer simply took the Thiokol statements at face value and assumed that there must have been more to the story, particularly behind the curtain of classified information.[10] This practice, too, would become common in the arena of personal flight. While developers were happy to exaggerate and over-promise, the press was more than happy to uncritically repeat almost anything it was told about the inventions.

*We must presume that no one had done any of these things because nowhere in any of the company's reports does it ever mention successfully doing any of these things. Interestingly, the company never even mentions *trying* to do any of these things. The most that Thiokol ever showed is one man broad-jumping, and his distance does not appear to match any of the company's claims.

Within the article about the solid-fuel jump belt, Thiokol revealed another experimental device to *Popular Science* readers: "The hush-hush flying belt." Thiokol provided the magazine with photos of a man wearing a couple of different contraptions. Photographs showed larger tanks strapped to a man's back with tubing extending from the tanks. In one picture, the exhaust pipes went straight out from the man's shoulders and then pointed downward. In the other, a single pipe snaked up and around the tanks and pointed down behind the man. The devices were clearly variations of Thomas Moore's Jet Vest.

> These hitherto unreleased photos show an early model of the secret "Flying Belt." In view at right, jets and control are folded for maximum convenience; in the other they are in flying position. Thrust angle is changed by moving the control forward or back, and amount of thrust or lift is varied by the movement of the hand grip. The belt would enable the wearer to rise quickly to an altitude of several hundred feet and fly "miles." How soon will the Flying Belt emerge from the laboratory? Reaction Motors' engineers estimate delivery of a workable one within two years.[11]

Oddly, Thiokol did not say if it had gotten the flying belt off the ground, and it did not mention how the unit worked or what fuel it would use. Yet people within Thiokol had figured out the answer: hydrogen peroxide.

Colonel Parkin followed these developments with great interest, remembering his own primitive experiments. Now, with major aerospace players involved in the field, he convinced TRECOM to investigate whether American scientists could build a practical "jump belt." Parkin would see to it that the scientists had the necessary funding to fully develop the technology.[12]

4

THE QUEST FOR THE ROCKET BELT

In 1959 TRECOM issued a Request for Proposal, asking for an authoritative report on the feasibility of "small rocket lift devices" to increase the mobility of individual soldiers. If the devices were considered practical, the government might fund research and development and even buy the finished product. Three companies sprang into action: Bell Aerosystems, Thiokol Corporation, and Aerojet-General all submitted bids.[1] Thiokol had already built its solid-fueled jump belt, and Bell, a helicopter manufacturer that also helped NASA with space technology, had been working on a flying belt of its own. In fact, Bell had made great strides by 1957, well before Thiokol showed its "Hush-hush flying belt" to *Popular Science*.[2] Bell just hadn't been bragging about it yet.

The army was not asking for anyone to build a device yet; they simply wanted a comprehensive report to explain what was possible with state-of-the-art science. Aerojet-General was granted the contract to conduct the study and draft the report. Interestingly, while the companies had all been working in this field and each felt it had made strides toward a practical device, the three were largely unaware of the activities of the others. As a result, each company believed it had a head start—but each company was veering off in a different direction.[3]

Aerojet-General built models and studied how best to launch a man into the air. The company investigated three different engine placements:

it could be in something that the pilot sat upon, it could be in a platform he stood upon, or it could be strapped to his back. The designers very quickly realized that for a machine like this to fly safely, the man being lifted would have to be able to control his flight through simple controls. And the simplest form of control would be to simply move his body. But would kinesthetic control work if the man was strapped to a rocket motor rather than standing on the device?

Aerojet-General tethered the models so they wouldn't crash and began testing them. The company immediately discovered that a platform with a seated pilot was bottom heavy and the pilot's movement was too limited for controlled flight. The standing version of the platform worked a little better but the backpack configuration worked best. After testing one such device and crunching numbers on its state-of-the-art IBM 704 computer, Aerojet-General determined the optimum location of the engine to be precisely eleven inches above the center of gravity of the man-machine combination.[4] The term "man-machine" pops up over and over throughout the report and it is apparent that any person flying the small rocket lift device envisioned by Aerojet-General was, indeed, becoming part of the craft. The operator's body would be integral to the structure of the device and the user's physical movements would be its controls.

The backpack device raised practical issues the Aerojet-General engineers addressed one at a time. When it wasn't in flight, a device that rode on the pilot would be a burden for the pilot to carry and hold on his back. Before taking off, the unit would be heavy with fuel. On landing, the pilot would have only his legs to use as landing gear. Aerojet-General determined that a backpack unit should actually sit on a girdle worn around the mid-section of the pilot, just above the hips. The bulk of the weight would then ride on the hips and legs and not just the back. The unit would have straps around the user's chest to hold the engines against the wearer's body. This configuration would reduce the likelihood of injury to the pilot, especially upon landing.[5]

Aerojet-General also studied the issue of fuel. Each type had its drawbacks. Some were more powerful but dangerous to handle. Others required mixing or were not as well studied. Aerojet-General confirmed that the most likely candidate for powering the small rocket lift device was hydrogen peroxide. Well known as a disinfectant or bleaching agent, the over-the-counter variety is usually only a 3 to 6 percent solution. A much higher concentration, in the range of 90 percent or more, makes a highly volatile rocket fuel. Scientists and engineers have known for years that when hydrogen peroxide comes into contact with a catalyst, it vaporizes quickly into steam. And many things will cause the reaction, including some metals and almost any organic material. The steam reaches 1,388 degrees Fahrenheit in two-tenths of a millisecond.[6] If this reaction happens in a small enough controlled space, the steam can be harnessed and directed, making it a powerful propellant. The beauty of hydrogen peroxide is that the exhaust, although very hot, consists entirely of steam and oxygen. The exhaust of the hydrogen peroxide–powered device would be clean and harmless to the atmosphere and environment.[7] People had been using hydrogen peroxide as a rocket fuel for years, and Aerojet-General believed enough was known about the fuel to make it a safe bet for testing.[8]

Perhaps the most intriguing aspect of the report was a claim about the skill level necessary to fly one of the small rocket lift devices. Even though the work was still largely theoretical, Aerojet-General had drawn the conclusion that the device would "allow an operator who has limited ground training and no previous free flight experience to safely perform a variety of maneuvers."[9] While Stanley Hiller had made similar claims in the mainstream media, it is unclear if anyone believed him. Here, an army contractor was making the statement authoritatively in a government-funded report.

In February 1960, Aerojet-General submitted its report to TRECOM entitled "Feasibility Study of Small-Rocket Lift Device," and concluded, "analytically a Small Rocket Lift Device is feasible. However, to establish complete feasibility and also determine the maximum extent of its usefulness, it will be necessary to build and test a demonstration model."[10] Aerojet-General was paid $56,456 for the report and hoped to be chosen to develop the demonstration model it had described.[11]

TRECOM then announced it was seeking proposals to actually build the small rocket lift device described in the feasibility study. Several companies submitted bids, offering to furnish the US military with a device that could help move soldiers around the Cold War battlefields of the

1950s. One of the major players was Thiokol, the New Jersey rocket fuel company that had proposed the jump belt that gave "the individual soldier a large increase in his ability to run, jump, skim bodies of water, and to permit him to fly."[12] In its 1959 report, Thiokol had proposed a variety of individual lift devices, each of which was "a man-mounted VTOL device."[13] There were three very different but related devices. The simplest, and the one for which the company would become known, was simply a belt that strapped a series of small solid-fuel rocket motors to a user's body. If the soldier—the report never conceived a civilian use for the solid-fuel rocket belt—needed a boost as he traveled the battlefield he could ignite one or more of the rocket motors, which would burn for a few seconds and augment his own walking, running, or jumping motion. Thiokol claimed that a soldier using one of these rocket units could cover a one-hundred-yard distance in less than ten seconds.[14]

Thiokol admitted its jump belt had limitations and in the same report suggested the answers: liquid-fueled or jet-powered flying belts. Thiokol told the government that it had examined the possibilities and come to the conclusion that a flying belt could be powered by a liquid fuel but its flight duration would be very short, perhaps no longer than thirty seconds. Thiokol noted that the short flight time would render the belt almost useless, since the cost and difficulty in operation could hardly be justified for thirty seconds of flying. If someone could build a small jet engine, however, a jet-powered flying belt could be constructed.[15] Thiokol looked around and did not find anyone making a jet engine small enough to put on a man's back. Even so, Thiokol called it a "unique but urgently needed lift device."[16] Thiokol spent most of its one-hundred-page report describing how the devices might work but spent almost no time explaining what urgent need was driving the company to design them. "The tactical mobility of the individual soldier in combat is still limited to running or jumping under his own power. With the limited nuclear war a reality, battle tactics will demand ever increasing soldier mobility of the type furnished only by Individual Lift Devices."[17]

Thiokol made working models of one of its designs, the jump belt. This posed a problem for Thiokol: how could it convince someone of the feasibility of these fantastic devices? Thiokol reiterated its claims that the jump belts would allow a soldier to run a hundred yards in Olympic record time, or jump across a fifty-foot gap. But had anyone actually done these things? Apparently not. Thiokol's report had diagrams that compared a soldier's one-hundred-yard dash times ("With Belt 9.3" versus "No Belt 17.8") and jumping potential ("With Belt 50" versus "No Belt 15") but the only photographs included were of soldiers wearing mockups of the jump belts.[18] For action, Thiokol provided drawings and photographs of models of the belts.

In response to the army's Request for Proposal, Aerojet-General, Bell, and Thiokol all submitted proposals to build small rocket lift devices. Thiokol's proposal was written by A. H. Bohr and contained much of the same information found in their earlier report on the feasibility of such a device. The introduction showed a simple drawing of a man wearing a primitive rocket belt with a set of nozzles at waist height. The unit had a single control, presumably to control the thrust, but no apparent controls regarding orientation or steering of the device.[19] This was not an oversight: Thiokol's engineers did not believe the unit, which the company now called a jumpBelt, could be made to fly. "The jumpBelt as proposed here is a small hydrogen peroxide rocket unit which can be worn by a human operator. Its potential users are to attain high running speeds, perform long leaps down, across, or up large obstacles, and to skim water at high speeds. Except for vertical jumping, the thrust force will not exceed the operator's weight and the device is essentially a ground borne unit."[20] One of the reasons Thiokol's engineers assumed the belt would have to be left on the ground is that the operator would need a free hand to be useful as a soldier. The engineers theorized that the operator would control the power of the unit with one hand and use his other hand to fire a gun or hold binoculars.[21] If the unit required both hands to operate, how could a soldier-operator do anything useful?

The proposed belt, if it worked, would give its operator a huge but short-term boost of power. If the operator was standing, it would assist the user in a jump. If the operator was running, the belt would provide a temporary push to add speed. Since this push was coming at waist level, Thiokol built a belt that fastened around the user's waist and had harnesses wrapped around the thighs. In some of the descriptions, the harness was described as a "modified diaper,"[22] which wouldn't have succeeded in making it an attractive product. Further, Thiokol listed several uses for the belt that it wanted to develop under the contract but the company was oddly silent on whether it had tried any of them yet. For instance, Thiokol suggested that a soldier could wear the Thiokol jumpBelt and skim across the surface of water, like a body surfer but faster, and on a level, wave-free body of water. Thiokol included drawings of these water-skimming soldiers, and the company proposed building a one-hundred-foot-long, five-foot-deep water trough just to test this idea.[23] That is, if the government gave Thiokol the money to start testing its device.

Fifty-five pages into the fifty-nine-page proposal, Thiokol threw in an afterthought. If the army wanted to see a small rocket lift device capable of actually flying, Thiokol could build one of those too. Thiokol engineers calculated that the flyBelt would only stay aloft for thirteen seconds, however. Thiokol thought so little of the idea that only four sentences were devoted to it. Still, "a system could be fabricated with very little additional drafting work."[24]

Bell's proposal pulled out all the stops and called for a hydrogen peroxide–fueled rocket belt that would fly. Bell had a leg up; the company had been working on the belt for a few years already and knew it would work. While the Aerojet-General report said such a belt was possible, and Thiokol had told *Popular Science* such a thing could be done in a few years, Bell had already been extensively rig testing one on a tether in Buffalo, New York, since December 1957.[25] Bell, led by an engineer named Wendell Moore, won the project and was promised $25,000 of government funding to build an operational prototype of what the company proposed.[26] Bell embarked on the small rocket lift device program on August 10, 1960.[27]

5

WENDELL F. MOORE AND THE BELL ROCKET BELT

Wendell F. Moore was an engineer at Bell Aerosystems, a company located near Buffalo, New York, and was not related to Thomas Moore, the man who worked on the ill-fated Jet Vest in Alabama. He came to work at Bell in 1945 after bouncing around at various manufacturers of automotive engines and aircraft accessories. He had studied aeronautical engineering at Kent State and Indiana Technical College, but doesn't appear to have finished his degree.[1] He was too busy working. Soon, he found himself at Bell Aerospace, working on cutting-edge space-age projects. Moore had a military-style short haircut and thick-rimmed glasses to make sure he looked the part of the hardworking engineer. He also had a landing strip in his backyard for the airplane he built in his garage.

He worked his way up through the ranks at Bell and found himself on an engineering team out to make history. They wanted to build the first aircraft to break the sound barrier in controlled, level flight.[2] The engineers planned to fly the Bell X-1 aircraft at an extreme altitude where the thin atmosphere would apply less drag on the plane. The lack of air in the upper atmosphere presented unique control problems. The plane's pilot needed to make very minor adjustments quickly, and in the thin air, normal control surfaces wouldn't do the trick.[3] Moore designed a system of small hydrogen peroxide–powered rockets mounted throughout the plane; they would fire when needed and push the plane, regardless of how thin the air was. And the rockets fired quickly. The system worked and the plane flew into the record books. The devices were referred to as reaction controls, and they would also be utilized on spacecraft, even as far along as the space shuttle.[4]

Moore first thought about the possibility of a hydrogen peroxide–powered flying belt in 1953 and talked his superiors at Bell into funding its development.[5] Over the years Moore told slightly different versions of the story but the gist was usually that he had sketched out a rudimentary rocket belt configuration, either on his kitchen table late one night or in the sand of the desert near Edwards Air Force Base while conducting X-1 tests.[6] In one telling, he stayed up all night at his kitchen table, excitedly drawing in the lines, tanks, and controls while his wife and children slept.[7] Bell knew about the army's interest in individual mobility devices and gave Moore the green light and a small budget with which to work.

It is unclear how much support Moore got from Bell in the beginning. Stories abound of nonbelievers within Bell and of Moore and his team scrounging parts because their initial budget was too skimpy for their purposes. Apparently, Moore could solve problems within the corporate culture as easily as he could the engineering problems he faced with his rocket belt. The man whose desk was next to Moore's told a writer, "He didn't get much support from management, but he kept plugging."[8]

When Bell won the small rocket lift device contract, Moore had already been working on his rocket belt for almost three years. He had not made a working model yet, but he had laid the groundwork for designing one. First, Moore designed and built a test rig, using nitrogen as the propellant. This device was not capable of untethered flight; it was a prototype with an umbilical cord feeding the compressed gas to the unit. The purpose was simply to design something a person could strap on and to demonstrate that if the unit delivered the necessary thrust to get off the ground, the operator could control the device and fly without getting hurt or killed. In many respects, this device was much like the one Thomas Moore had built and tested years earlier.

The rig Wendell Moore built to test the compressed nitrogen rocket belt had mounted pipes on a harness worn by an operator; the pipes took compressed nitrogen fed from a line and split it so that it then blasted out both sides, behind the operator's head. The operator hung in the air on tethers from the rig and held onto handlebars attached to the harness. Moore hoped that the operator could move the handlebars to balance while the nitrogen was blasting through the rig. Another engineer—not the operator—controlled the pressure of the nitrogen. Since the device was tethered, it allowed the nitrogen to be fed into the unit without cumbersome tanks on the back of the operator.

Moore was not content to simply build his inventions and let others test them. He personally tested the unit, although he let a couple of other test pilots at Bell try the contraption later. Moore made his first nitrogen-powered, tethered flight wearing a rocket belt on December 17, 1957.[9] No one was certain that the device was safe; every time Moore fired the belt there was a doctor standing by, just in case something went wrong.[10]

The first tests weren't promising; Moore floundered at first. Bob Roach, a project engineer who specialized in explosives and designed rocket motors for Bell, watched the earliest tests.[11] He knew it all made sense on paper, but he couldn't help but notice how wildly different the results were for different operators. Jim Powell, a flight research engineer for Bell, strapped himself into the rig and gave it a try. To him, the device seemed unstable but showed promise. When he pushed and pulled on the control bars, the device went the way he was hoping it would. If they could get a more responsive throttle and let the pilot control it, perhaps the kinesthetic control was the key.

One man watching the tests was also filming them. Tom Lennon was Bell's official company photographer. He asked if he could give the test rig a whirl. If he was volunteering and had just seen how the rig worked, why not? Moore allowed it. Lennon climbed into the harness and told the engineers to give the unit power.

> Lennon gave the "thumbs-up" signal, and away he went—five, ten, almost 15 feet into the air—up, then down; up again, down and up for the next three minutes without the slightest hint of instability. This experience, with different operators with the same equipment and identical test conditions, indicated that even though a

> stable configuration might be forthcoming in the actual rocket belt designs, the psychological attitude of the operator himself would play a significant role in success or failure of the test.[12]

When Lennon had been filming Moore, he thought the inventor had less control when he didn't keep his feet together. When Lennon tried the belt, "I pinned my feet together and had no trouble controlling it."[13]

Moore drafted a report that Bell gave to the army: the proposed rocket belt would be stable and subject to kinesthetic control. The second phase would be to cut the umbilical cord and put the fuel supply on the back of the operator.

Moore submitted a patent application for his rocket belt design on June 10, 1960. Moore's idea was ingenious and simple. It consisted of three tanks that rode on the back of the pilot. Two of them were filled with hydrogen peroxide and one was filled with nitrogen. The compressed nitrogen did nothing more than force the hydrogen peroxide out of the tanks through a throttle valve and into a gas generator assembly. There, hydrogen peroxide washed over a silver mesh catalyst, which caused the chemical reaction and drove the belt. The hydrogen peroxide burst instantly into 1,300 degree steam and expanded so rapidly that when the reaction was vented through nozzles, it created quite a bit of force. At the exhaust tips, the steam was traveling one thousand meters a second.[14] It had been enough to steer the X-1 at supersonic speeds and it would prove strong enough to lift a man from the ground. And on August 10, 1960, Bell received the funding to prove this when the army awarded the contract for the small rocket lift device program.[15] Moore had already figured out the configuration he would use for his rocket belt.

The heart of the belt was the throttle valve and the gas generator assembly. The throttle valve controlled how much hydrogen peroxide flowed into the gas generator, where the fuel reacted with the catalyst. This valve would later become the focus of would-be rocket belt builders everywhere. It was not an easy device to manufacture and its complexity stood in stark contrast to the rest of the belt, which appeared simple. A person also could not discern how the valve was constructed by simply seeing it. Many people would photograph and study the Bell Rocket

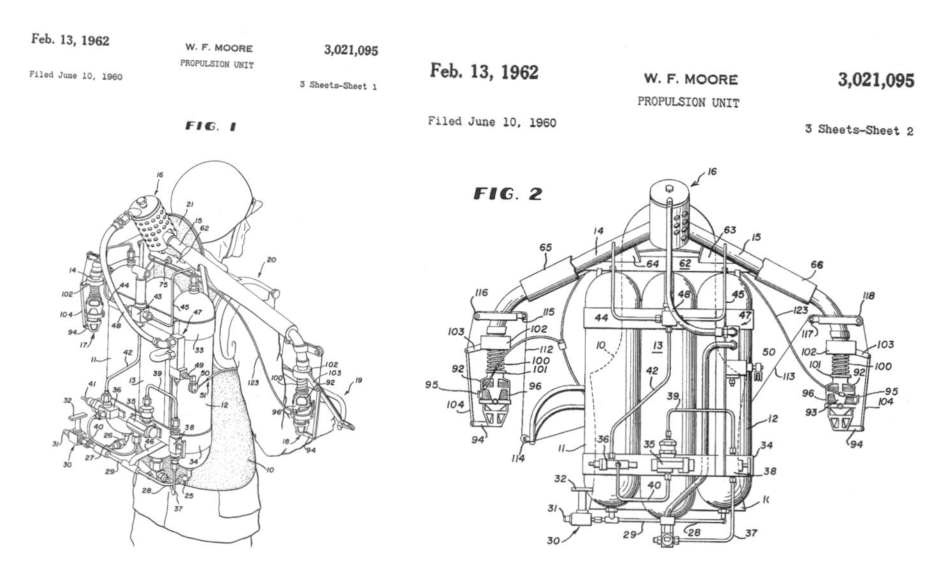

The rocket belt patent filed in 1960 by Wendell Moore laid out a design that would remain in use, largely unchanged, for the next fifty years. *Author's collection, Courtesy of Bell Helicopter Textron*

Belts hoping to find out the secret of the valve. The throttle valve was designed and assembled by "the National Water Lift Company of Kalamazoo, Michigan."[16] It "resembl[ed] the one designed for use in the Mercury Capsule manual reaction-control system."[17]

There was little room for error, considering how volatile the fuel was. The entire valve was made from aluminum and consisted of a plunger inside a sleeve. National Water Lift valve designers Vic Heine and Gene Belco determined the valve needed to be made from hard anodized aluminum, which is "much harder than steel." The components of the valve had to be made by outside vendors because the hard anodized aluminum required specialized machine work and "special diamond grinding tools." The clearances inside the valve were so critical that some of the valve's manufacturing tolerances were measured in millionths of an inch.[18] The small clearances meant that operators of the belt had to be extremely careful about the quality of their fuel and nitrogen. Any impurity could cause the valve to jam. And because the valve was part of a rocket motor that could get hot very quickly, the entire valve had to be made from the same material. Otherwise, parts made of different metals might expand at different rates and cause the valve to fail. Each Bell Rocket Belt featured the valve, stamped "S.R.L.D." and "National Water Lift Co, Kalamazoo, Mich."[19]

The gas generator was made by Bell. At this point in time, hydrogen peroxide had been in such wide use that in the reports, Moore referred to the hydrogen peroxide gas generator simply as a conventional design.[20]

When fired, the rocket motor was deafening. The noise level of the exhaust was in the range of 150 decibels, "sufficient to cause strong discomfort to unprotected ears."[21] Bill Suitor, a rocket belt pilot, likened it to the same intensity as the noise from a jackhammer.[22] People who saw the belt fly almost always recounted the noise. Many were almost as awed by the sound as by the sight. The noise issue would dog the flying belts to some extent throughout the program. Critics always cited it as a handicap, and even the earliest scientists recognized it as a problem.[23] By 1958, Aerojet-General had experimented with variously shaped exhaust nozzles to try to reduce the noise level. No matter what the company tried, the rockets were incredibly loud. The engineers eventually settled on earplugs as the only practical solution.

The rocket belt's frame was a fiberglass corset shaped to fit the pilot. A plaster cast was made of Wendell Moore's back and a mold was made to replicate his shape. The mold was built up an inch to allow for foam padding, and then fiberglass was laid over the mold. Even the padding was subjected to rigorous testing. Bell found three candidate materials and soaked them in hydrogen peroxide to see how each foam would interact with the rocket fuel. One, Dow's Ethafoam, did not react at all to the rocket fuel and was chosen for the belt.[24] Once the fiberglass had hardened overnight, an inch of the Ethafoam padding was added to the corset and Moore had a custom-fit fiberglass harness to attach the rocket components to his back.[25] The tanks attached to the corset and the weight of the tanks and the controls rested largely on the hips of the operator.[26] There were straps and belts tying it all together. Bell also designed a special flight suit for the rocket belt pilot, made from a polyvinyl-impregnated cloth that was the best material to be wearing when working with hydrogen peroxide.[27]

To be safe, however, the first time Moore and his team test fired the hydrogen peroxide motors they did it with the rocket belt strapped to a plaster dummy in a test cell.[28] It was October 14, 1960. They had pressure tested the tanks and checked all the fittings and valves. They filled the respective tanks with fuel and nitrogen and planned to fire the engines until the hydrogen peroxide ran out. They had set up the rocket belt with remote control linkage so the engineers could step back a distance when they fired the motor—just in case.[29] Moore hit the throttle and the rocket belt put out so much thrust it twisted in its mounting bracket. When it twisted, the exhaust nozzles no longer aligned with ducts that vented the steam out of the test cell. Instead, after just seven seconds, Moore could not see what was happening inside the test cell. He killed the throttle and hoped the rocket belt wasn't harmed. It was fine and they reinforced the mounting hardware.

From this point forward, all of the test cell firings went perfectly.[30] The belt was fueled and fired, measurements were taken, data recorded, and the belt would be checked to see what effects the firings had on it. Moore assigned other engineers to handle these tests as Bell needed to run quite a few and Moore had other issues to resolve. On October 18, 1960, the belt was tested on the second shift. An engineer, noted as H. M. Graham, conducted the test. Harold "Hal" M. Graham was a young engineer who would conduct four of these test cell runs. Shortly after, he would feel burned out from working the second shift at Bell and quit.

Test cell firings allowed Bell engineers to examine the belt and make improvements when needed. One question Moore decided to answer definitively was, How loud was the rocket belt? He measured the volume of the sound by placing a meter where the pilot's head would be and got a reading of 131 decibels. The volume measured between the exhaust nozzles read 133.5.[31] The belt had been configured with a squeeze handle for a throttle but the engineers decided a motorcycle twist grip would work better.[32] Bell engineers purchased a Harley Davidson throttle assembly for testing.[33] At this point, Moore was simply trying to prove that Bell could make a viable rocket belt; as a result, some corners were cut. In its initial report on the development, Bell stated that many off-the-shelf parts had been used in the prototype belt because these parts were handier and cheaper. Later versions of the belt could be more refined and lighter. While weight would later become very important, it was not the primary consideration at this stage of development. As Moore tested the belt, it weighed 79.57 pounds without fuel.[34] Later belts weighed 63 pounds empty, and a full load of 47 pounds of fuel brought the unit's take-off weight to 110 pounds.[35] This is not to say the belt wasn't high tech. Some of the parts used on the belt, such as the nitrogen shut-off valve and a high-pressure, ten-micron filter, were developed for the Mercury space program.[36] The extremely fine filter was necessary because of the small tolerances inside the throttle valve.

Moore also timed how long the rocket would run with full fuel tanks. While he had been hoping for a time of thirty seconds, the best he could manage under real-world conditions was twenty-two seconds.[37] Moore and the others had always known that the flight limitations of the belt would fall somewhere around this range. They hoped to prove that the device would fly and could be controlled; then they would work on the duration issue. Control turned out to be the easy problem; longer flight times would prove elusive. "One of the objects after building it was to extend the range and look at other propellants. But the problem is that the more powerful you get, the higher the risk you take," Bob Roach said decades later.[38] Some of the scientists were concerned, but they had all encountered and solved engineering problems before. They marched on.

After eighty-eight successful test firings without destroying the dummy or the test cell, Moore drafted a status report for Bell to submit.[39] He noted that the tests went perfectly, and Bell was ready for "hot-firing" the rocket belt. It wanted to move to manned flights, both tethered and free.[40] The army read the report and gave the go-ahead.

The scientists and engineers at Bell had worked out a host of technical issues regarding hydrogen peroxide. They had used the fuel before but the rocket belt burned more of it, and strapping the device to a man's back raised the stakes substantially. Hydrogen peroxide was extremely touchy: it would react when it came in contact with seemingly anything. It burst into steam when it contacted the silver in the catalyst bed. Spilled on wood, it would start a fire. Even putting it into the tanks of the rocket belt was no easy task. If they simply found a tank and started putting fuel into it, they might discover—the hard way—that there were impurities inside the tank that would react with the hydrogen peroxide.

The answer lay in "pickling" the tanks. The tanks had to be cleaned thoroughly and then filled slowly with low-grade hydrogen peroxide. The lower-grade fuel would still react with any impurities, but to a lesser extent than the high-grade fuel. Each reaction would remove some impurities. The tanks would be drained and the process would be repeated with a higher concentration of fuel. The process might take days, but the tanks were eventually free of impurities.

The rocket belt controls went through an evolution also. At first, Moore and the others thought it might be necessary for a pilot to have a full set of controls to cover not only the throttle but the pitch, roll, and yaw—all directional changes made with an airplane. There was a problem, though, with how a pilot would work such controls while flying the rocket belt. Moore considered having the throttle and yaw controlled by a right-side handle and the pitch and roll controlled through a handle on the left. At one point, he even considered connecting the pilot's helmet so that he could steer the rocket belt in flight by turning his head.[41]

For the most part, the rocket belt could be controlled kinesthetically, using in addition a set of motorcycle-style handlebars for the pilot to hang onto and control. One handle was a throttle and the other controlled the belt's yaw. The pilot would move forward by leaning forward and pulling on the handlebars. He would slow down and stop by leaning backward and pushing the bars forward, away from himself. While flying, "leaning" was accomplished by pushing and pulling on the handlebars of the rocket belt. The yaw could not be controlled kinesthetically. If a rocket belt pilot wanted to turn around in flight without flying in a huge circle, there needed to be some way for the rocket belt to rotate on its vertical axis. Moore experimented with a few different methods and wound up with "jetavators." These controls were connected to the left-hand twist-grip of the pilot and caused changes in direction of the exhaust at the tips of the nozzles. When the control was moved, it made the jetavators turn in opposite directions; when one went forward, the other went backward. This caused the pilot to turn in place, spinning one way or the other.[42] At first, Moore had thought it might be necessary to be able to swivel the exhausts a full 360 degrees—with an allowance made so the exhaust would not point in toward the pilot and burn him—but testing proved this was unnecessary.[43]

Wendell Moore then tested the rocket belt with a live person: himself. On December 29, 1960, he strapped it. He attached himself to tethers to keep from being launched or slammed out of control, and he fired the engine. Moore was an engineer and not a test pilot. Bell had sold the government not only on the feasibility of a rocket belt, but on one that could be built that "would be safe for operation by relatively inexperienced personnel."[44] This "flying machine for the common man" theme had been a sales pitch of the Hiller Flying Platforms and would remain part of the equation with the flying belts.

Each time the belt was flown, whether on a tether or in free flight, at least ten people were present, counting the operator of the rocket belt. Along with the doctor, there were two photographers to make sure that everything was as well documented as possible. One was Tom Lennon, the man who had successfully flown the early test rig device powered by compressed nitrogen. Lennon spent twenty-eight years photographing almost everything done at Bell, whether he was riding in a chase jet and filming the X-1, recording a hovercraft in Greenland, or standing in a huge hangar in New York, snapping pictures and shooting film as Wendell Moore careened around on the end of a tether. Rounding out the audience were engineers who specialized in testing, human factors, and other specialties, men holding the tethers, and others who fueled the tanks. Moore strapped himself into the rocket belt and ran through the pre-flight checklist. Sure that everything was set, the cameras rolled and Moore took flight.[45] The tethered flight lasted less than ten seconds but it proved the rocket belt put out enough power to get a man off the ground. It was the first successful tethered flight of a Bell Rocket Belt powered by a hydrogen peroxide motor.

"Several things were immediately apparent. One was that the building itself was much too small for this type of test."[46] Moore moved the testing outdoors and configured a tether rig, but as temperatures dropped, the exhaust from the device became so dense that it obscured the engineers' views. During one test, Moore landed badly and fell backward, denting the bottom of the rocket belt and causing a slight leak. The engineers quickly built a tank guard to prevent such a problem in the future, and they moved the testing back indoors. They commandeered a "large experimental flight hangar" where they would have all the room they needed to fire the rocket belt without hitting things in the process.[47]

As Moore discovered problems with the design of the belt during the tests, he fine-tuned specific aspects. He confirmed that the Harley Davidson twist grip worked better than the squeeze throttle.[48] The exhaust pipes were originally aimed straight down, and the jet blast hit Moore's feet and made him unsteady. The engineers repositioned the tips so that the exhaust blasted five degrees away from the vertical. This small change made a big difference in the comfort of the operator and also made the belt more stable.[49]

Moore confirmed that the tanks on the unit only provided enough fuel to run for twenty-one seconds. After that, the rocket belt would stop lifting. If Moore was off the ground, he would fall unless he was caught by the tethers. The unit could easily lift him from the ground. But once airborne, Moore found it tricky to control the device. As soon as he felt like he was gaining control, he'd be out of fuel. If he overcorrected, he'd be blasted sideways. An engineer who worked with Moore later told a writer, "He whacked himself against the walls a couple of times, and everyone was saying, 'What the hell is he trying to do?'"[50] Moore kept at it, firing the rocket belt with the safety lines still attached, in his attempts to get a little bit closer to free flight. The comments he recorded for one tethered flight described his "legs flailing in a pendulous fashion." On another, his "motions were erratic."[51]

On February 17, 1961, Moore suffered his first major setback. During his twentieth tethered flight, Moore was trying to hover and he was unaware that one of the tethers holding him above the ground had been scraping back and forth over a piece of metal near the ceiling of the hangar. As a result, the tether could not hold his weight when the rocket motor stopped firing. He fell onto the cement floor from eight feet, shattering his kneecap.[52] Moore glossed over the accident in his report on the device and only wrote a few sentences on the incident, simply calling it a "knee injury." Surprisingly, the fragile parts of the rocket belt hadn't been damaged in the fall; the engineers realized it was a good idea to put a guard over the valve assembly at the top of the unit to prevent damage.[53] But Bell decided that Moore should not be test flying his rocket belt anymore. He wound up with a cast on his leg from his foot to his thigh, and he wouldn't have been able to fly until the cast came off.

6

THE ROCKET BELT FLIES

Unlike the Wright Brothers, Wendell Moore would not be able to personally fly his own invention, and this upset him greatly. His daughter later told a writer, "It broke his heart when he found out he couldn't fly it anymore, because that was his baby. It really did a number on him." Bob Roach, the project engineer on the rocket belt who sat next to Moore, later said, "I think that was his greatest disappointment, that he never got to make a free flight."[1]

Bell had recently rehired Harold "Hal" Graham, the engineer who had been working second shift but quit. He had worked on a variety of engineering tasks at Bell, including work with the rocket belt and some of its components, but now he was hired to be the rocket belt pilot.[2] It is unclear if Bell simply wanted to have someone else trained as a pilot, or if the company was beginning to worry about Moore's safety. After all, what would happen to the program if the rocket belt killed the man developing it? The company rehired Graham just two days before Moore's accident. The timing of the hire suggests that someone in management may have played a very good hunch. Moore liked the fact that Graham knew engineering, but also that he was athletic. He had played hockey, and Moore believed the balance and strength required for the game would help a pilot learn to fly the belt. Graham had a degree from Rensselaer Polytechnic but no experience piloting anything other than an automobile.[3]

Graham's new job meant he would work the day shift, and he later said that one of the reasons he returned to Bell was that he would go from being just a small part of a program—one of thousands of engineers—to being at the center of the action. It didn't matter how many people worked on the rocket belt: Graham would be the one flying it.[4]

Being the center of attention is one thing; learning to fly a rocket belt is another. Those who have mastered it often point out how hard a task it is, regardless of how fun or easy it looks to spectators. One pilot likened it to learning to stand on top of a basketball and then, once you have gotten your balance enough to stand there without falling off, to be able to walk around a gym floor, atop the basketball, without losing your balance. "You could take a basketball out on the basketball floor this afternoon and try to stand up on it and you'd convince yourself, 'That's not possible. You can't do that; it just won't work.' Until you see a clown go out and do it and make a fool out of you by doing all kinds of acts, balancing himself on a basketball."[5]

Moore brought Graham in on March 1, 1961, strapped him into the rocket belt, and connected the tether, which had been beefed up since Moore's accident. They fired the rocket belt, and Graham lifted off. Afterward, he told the flight doctor it felt like a hook had picked him up from behind.[6] Graham flew on a tether once a day and after each flight, Moore filled out a report card of Graham's performance. The performance evaluation asked, "What was the Flight Operator attempting to do? Please sketch his flight plan as you understood it. What was his actual performance? Please sketch it below." Moore then graded the flight for accuracy and flight characteristics on a scale of one to five, and he filled out a comment section. If need be, he could continue his comments onto the back of the report.[7]

While experimenting with the yaw controls, the observers saw some weird things happen. On three different occasions, the rocket belt pilot initiated the yaw control only to have the belt keep turning the pilot after he had tried to stop rotating. In each of those three events, the pilot solved the problem by simply landing. The amount of turning force from the yaw controls was not too much to keep the pilot from landing safely.[8] Moore encountered various issues such as this along the way; he resolved them and continued the testing.

It took Graham a while but after he had done thirty-six tethered flights, Moore thought Graham was ready to fly without the tethers.[9] On April 20, 1961, they took the rocket belt out to the Niagara Falls Airport. It was a chilly day, with the temperature flirting with the freezing point at 35 degrees.[10] They marked a one-hundred-foot strip in the grass "near the threshold of Runway 32" for Graham to fly over. They chose a grassy spot so that Graham would have a more forgiving surface to work on. They drew a target at each end of the strip. Graham put on the belt and stood on one target. After completing a preflight checklist, it was time. Graham withdrew the safety locking pin from the throttle control and handed it to the project engineer, who gave him the go-ahead and stepped back to watch.

Graham squeezed the throttle for a quick burst, just to test the system. Then, he gunned it.[11] With a huge vapor cloud trailing behind him, he lifted off and traveled the length of the strip.[12] With that, Graham became the first person to fly a rocket belt untethered. His flight lasted thirteen seconds and covered 112 feet.[13] His entire flight had been conducted with his feet no more than eighteen inches from the ground.[14] After he landed, the men noticed that on the nearby road, cars had come to a halt when they saw the man blasting off in a cloud of steam. How many of them knew they had just become "eyewitnesses to a bit of history"?[15]

It had also been witnessed by Dr. F. Tyler Kelly, the Bell doctor who was present at every flight of the rocket belt through Phase II of its testing. Before and after each flight, Kelly recorded the blood pressure, respiration, pulse, and other pertinent data of the pilot. He also questioned the pilots before and after each flight to determine their subjective feelings and apprehensions about the flight, and recorded the "description of sensations in flight."[16] Moore noted on the flight report card, "The operator felt a high-level confidence in his control."[17] Bob Roach was out of town that day and missed the flight, although he saw many of the flights in the days to come.

Graham, Moore, Kelly, and the rest of the flight team moved the next flights to Niagara Frontier Golf Course so Graham could practice on rolling terrain. They were also concerned that the airport runway had been too visible to cars driving by on a nearby public road.[18] After a couple of free flights, Graham suggested it might be helpful if he wore goggles during his flight. He tried them and they helped, but they got foggy. Bell had the goggles steam-proofed and they worked perfectly from that point forward.[19] The noise generated by the rocket belt concerned Bell engineers as much as it had concerned everyone else who had studied hydrogen peroxide–powered flight. On the twentieth flight, they measured the sound level at 125 decibels.[20]

After the successful Niagara Falls Airport flight, Moore rewrote some of the figures for the rocket belt. There had been some minor changes to the belt since the idea had first been pitched to the military. For instance, there were now two safety guards on the belt, one covering the bottom of the tanks and one covering the valve assembly. The throttle had also been reworked. As a result, the rocket belt now weighed seventy-eight pounds without fuel. Hal Graham weighed 173 pounds fully suited for flight, and the fuel weighed a tad over fifty pounds before launch. According to Moore's calculations, the rocket belt still had fifty pounds of lift to spare; if the operator had a way to hang onto or carry a payload while flying, it was technically possible.[21]

By May 25, 1961, Graham had successfully flown the rocket belt twenty-eight times. He had flown slalom courses, crossed rivers, and jumped fire trucks with it. They declared the "development flight testing" completed.[22] Moore tore the belt down and inspected it to see what kind of wear and tear it had suffered from all of its flights. Surprisingly, the belt appeared to be no worse for the wear. One small o-ring had been replaced during testing and it was operating perfectly now. Even the catalyst bed, which reacted with the hydrogen peroxide, showed no noticeable wear from all the firings. It had been subject to 199 runs, 171 of which had been bench firings.[23] Still, the belt ran out of fuel at twenty-one seconds; all of Graham's flight time at this point added up to less than ten minutes in the air. Yet Moore felt the rocket belt could be unveiled to the public.[24]

For the reports provided to the army, Moore used data from his own flights and from Graham's, even though only Graham had free-flown the rocket belt. Moore reported to the army that there were only two fears and apprehensions experienced by the rocket belt pilots. The first was the fear of accidentally cutting the throttle while still airborne. Graham had done that a few times and when he was too low, he hit the ground before he realized his mistake. To solve this problem, Bell engineers added a built-in resistance to the hand grip when the throttle moved below 70 percent. The other fear was of running out of fuel while still in the air. With only twenty-one seconds of flight time, the rocket belt was running dangerously low on fuel the moment it took off. The pilots of the rocket belts learned to accomplish as much as possible in those twenty-one seconds, however. Moore experimented with different warning devices set by timers to warn pilots when they had only fifteen or ten seconds left of fuel. A warning light was tried but Moore worried that something might obscure the pilot's view of it. A warning buzzer was tried but the pilot couldn't hear it over the roar of the engine. He settled on a vibration device installed in the back of the operator's helmet. It pressed up against the pilot's head so that he could feel it when it began to vibrate. After fifteen seconds it vibrated periodically; after twenty it became constant.[25] Of course, at twenty-one seconds, the rocket belt was out of fuel and would suffer from "lift degradation."

Moore also noted for the army that the only real issues of operator safety encountered had been bumps and bruises that could be avoided by simple padding. As a result, pilots were fitted with protective devices for the elbows, shins, and knees, and Bell even designed and tested a "metallic groin protector."[26]

During the Phase II operation of the rocket belt development, Bell had done some market research and asked military personnel their thoughts on the small rocket lift device and whether they might consider such a thing useful. Clearly, Bell was hoping to get some insight into how to pitch the rocket belt to the military when it was done fine-tuning it. The answers the company got weren't encouraging, however. The US Army quartermaster said the unit sounded too heavy. It was too loud. The corset was probably not capable of withstanding sudden acceleration, and the blast from the exhaust nozzles was probably dangerous. What would happen to a rocket belt operator who ran out of fuel in the air? Training would probably be difficult. But the solicited comments ended on a high note. The project sounded intriguing. "With these limitations in mind, it would still constitute a powerful military tool."[27] Other departments weighed in with similar comments, although the US Army Aviation Human Research Unit added one small item to their wish list. Could Bell make a device that carried thirty minutes worth of fuel?[28]

Wendell Moore submitted his official report on the rocket belt in July 1961.[29] In it, he explained how Bell had developed the belt and had proved it could be flown safely. It was still unclear how the device could be used with the flight duration limitation. Moore wrote in his conclusions that "hydrogen peroxide, although very successfully utilized on this program, would have a limited tactical use due to its handling characteristics and limitations at low ambient temperatures. A better tactical propellant must be found."[30] The army would have to think about that some more. It accepted the results of Bell's work and declared the contract terminated.

On June 8, 1961, before the contract was terminated, Moore and a contingent from Bell brought the rocket belt to Fort Eustis and demonstrated it by having Graham fly over a "deuce-and-a-half" army truck parked in front of the spectators. The flight lasted only fourteen seconds and Graham traveled only 150 feet but it made quite an impression. The *New York Times* reported on the existence of the Bell Rocket Belt the next day. The *Times* called the flight "short but spectacular." It was a "significant demonstration of how soldiers might move around a battlefield of the future."[31] Bob Roach, who had designed many of the rocket motor's components, was there as well.[32] Graham also flew over a helicopter parked on the ground.[33] A week later, he demonstrated the rocket belt at the Pentagon, where three thousand office workers spilled out of the building to watch the real-life rocket man do a flight over an army staff car conveniently parked on the lawn.[34] A short while later, he was invited to appear on the television show *To Tell the Truth*. Celebrity judges quizzed Graham and two actors to determine who was the real rocket belt pilot they had been hearing so much about lately.[35]

From this point forward, tales of the rocket belt found their way into the mainstream media on a large scale. Bell found that journalists loved watching the flying belt in action and never ran out of superlatives to describe the flights. *Saturday Review* for July 1, 1961, ran a full-page feature: "Man Learns to Fly in a Steam-Powered Corset," and said the device was powered by a steam engine.[36] The rocket belt did generate steam, but Bell would not have used that terminology. The article introduced the readers to Hal Graham, the only man "known to have flown in this device."

Moore was forthcoming with information on the rocket belt when speaking to the press, unlike Thiokol engineers who had claimed their work was too classified to be described in detail. Moore told the *Saturday Review* that his rocket belt had been built using off-the-shelf parts simply to prove it was feasible. If he was given the go-ahead, he could make a rocket belt out of lighter materials and with a little more investment could tweak longer flight times out of it. So far, Graham had flown the belt more than thirty times and his longest flight had been 360 feet. He had rarely flown higher than three or four feet, but he had flown up and down hills. Moore said that none of these numbers indicated limitations of the belt. The article ended with a statement that the belt weighed more than a hundred pounds and managed speeds of thirty miles per hour, but that the weight could be lowered and the speed could be increased. Even better, Moore told them that while the current cost of building a rocket belt was unknown, "it could come down well below the price of an automobile."[37] The makers of the rocket belt had joined the tradition of overselling an individual lift device.

It is important to remember that the twenty-one-second flight limitation for the Bell Rocket Belt never changed. Reporters would sometimes mention it in passing or bury the fact at the end of an article, but it was usually overshadowed by glowing praise for the belt. The press coverage seeped into a variety of outlets. *MAD* magazine parodied the belt in its December 1961 edition in "Amazing military rocket-belt developed: Army unsure of practical use."

> Recently, the American public was startled to see movies and news photos of the successful testing of a perfected rocket belt. The pictures clearly showed a test engineer being propelled over land, water, trees and trucks at a height of fifteen feet. However, the Army confessed that it had no ideas as to the practical application of this ingenious invention. And so, with this article, *MAD*, in its typical public-spirited way, offers some suggestions.

At the time, poking fun at the army was a major theme of *MAD* magazine, and they then spent four pages illustrating, with cartoon drawings, things GIs could do with their rocket belts. "After wild night in town, sleepy GI can sleep while standing at attention by using rocket belt at half-power." Some of the suggestions were probably closer to the army's vision than *MAD* might have guessed. One cartoon showed soldiers landing on a beach wearing rocket belts, but gave a comical reason for doing so: to avoid the problems on landing craft, like "overcrowding, pushing, shoving and B.O." Likewise, *MAD* suggested that military surplus sales might result in the belts being worn by commuters and used in agriculture and construction.

To a rocket belt enthusiast, one of the funniest things about the article is a drawing of the rocket belt, as imagined by the *MAD* artist. While including absurd accessories like a Kleenex dispenser and a cigarette lighter, it also included a "low fuel warning buzzer" and a "no fuel warning buzzer."[38]

Hal Graham went to Fort Bragg in October 1961 to participate in a combat readiness demonstration for President John F. Kennedy. Kennedy sat and watched as members of the 82nd Airborne Division drove military machinery around; at one point an amphibious vehicle trolled by in a small pond in the middle of the demonstration area. Graham was on the boat, wearing a fully fueled rocket belt. As the boat drew even with the president, Graham blasted off and zoomed across the water toward the presidential entourage. The blast of the rocket belt kicked up a cloud of spray in the water below him. He performed a perfect landing directly in front of the president and snapped off a stiff salute, captured by *Life* photographers.[39] Kennedy returned the salute, "in fine Navy style."[40] Without question, it was an awe-inspiring moment. The *Buffalo Evening News* covered the event and found an army officer who watched the president as Graham landed. He said Kennedy's expression was "wide eyed and open mouthed, just like a kid."[41]

After the first demonstrations were so widely hailed in the press, requests began pouring in to Bell for more rocket belt demonstrations. At this point, Bell was still pitching the rocket belt to the US Army, hoping for further development funding. Bell honored requests only after getting approval from the army. Even so, most of the requests had to be turned down for "economic reasons." Bell sorted through the letters and found them to range from "totally ridiculous to criminally suspicious. One gentleman demanded his own rocket-belt so that he could put his hands on a $1,000,000 treasure which, he said, could not otherwise be reached." The request was denied along with most of the others.[42]

In December 1961, *Popular Science* once again told its readers of the exciting advances in personal flight. But instead of pictures of mere mockups of rocket packs, Bell provided the magazine with photos of Hal Graham hovering five or six feet above the ground with his back—and the rocket belt—to the camera. While the rocket belt had been developed for military use, Bell clearly thought it would be helpful if the American public thought the rocket belt might be another appliance to make life easier. Reading between the lines of the article "This Man Can Broad-Jump 368 Feet," the message comes through loud and clear: "It's easy." The design was "utter simplicity." They would be available for civilian use "within two or three years" and would soon "prove as versatile as helicopters."[43] The news coverage also shifted the focus of attention from Moore, who had turned the belt into a reality, to Graham, who was always the one flying it when the press was around.

The *Popular Science* sales pitch was sealed by Graham's demonstration for the magazine. Bell let the reporters watch a rocket belt flight up close. Graham explained how the unit operated as he put the belt on and told a story of how his flights had frightened and amazed local residents. A waitress from a local diner had supposedly been driving home bleary-eyed after an all-night shift when she heard a roar coming from the Bell property. When she looked over, she saw Graham rising up on a trail of steam. The sight of the hovering man, wearing a helmet and goggles, so mesmerized the woman she temporarily forgot she was driving a car. She ended up in the ditch but was not hurt. The way Graham told it, she was more shaken up by seeing the free-flying man than she was by running her car off the road. After that, Graham told the reporter, Bell vowed to stop testing the unit so close to public roadways.* Graham then motioned for the writer to step back.

---

*This story is reminiscent of Roach's about the first test flight. However, Graham was not wearing goggles on his first flight. It is likely he was exaggerating a bit here for dramatic purposes. "This Man Can Broad-Jump 368 Feet," *Popular Science*, December 1961, 105.

> A twist of his right hand on the fuel control shot a thin stream of nitrogen-pressurized [hydrogen] peroxide through the silver screens. Superheated steam blasted the ground, stirring up dust like a miniature tornado. Graham opened to full power and slowly rose into the air, twin jets shrieking. He sailed up and over a truck, and landed as lightly as a dandelion seed on the other side.[44]

Graham explained kinesthetic control and how most of his maneuvering in the air was simply the result of body movements. To go forward, he leaned forward. To stop or slow down, he leaned back. *Popular Science* was sold on the idea. Graham mentioned the high cost of the rocket belt's operation, about one hundred dollars per flight, and the short duration. The writer said the flights were limited to twenty-five seconds, but the flight the writer witnessed was shorter than that. The limitations of the rocket belt were buried in hyperbole. The article ended on a high note: "And when we've colonized the moon, rocket belts would offer an ideal way to get about on its craggy surface."[45]

Graham was becoming quite a star but found it all a bit humbling as well. People broke into spontaneous applause when he landed and he had to remind himself that they were clapping for *him*. "After every flight, people are applauding. The reporters are there. The photographers, and I'm thinking, 'Oh, they're looking at me.'"[46]

Whenever Graham traveled with the rocket belt, the support team often included as many as twenty people, some of whom were pilots. It inspired Graham to get his own private pilot's license in 1962. He bought a small plane, and whenever he wasn't flying the rocket belt he was flying his single-engine Piper.[47] It was a bit quieter and stayed in the air longer.

While Graham's *Popular Science* article was on newsstands, Bell began work on Phase III of the small rocket lift device program: further refinement of the rocket belt.[48] The new research was aimed in two directions: to make the unit safer and to make it stay in the air longer.

Moore and his team decided to revisit the warning buzzer in the helmet. It wasn't actually measuring the fuel levels in the tanks, it was simply operated by a timer. Moore experimented with different sensors to tie the warning system to the actual fuel level—sort of like how a gas gauge in a car works—and also tried different ways of warning the pilot about low fuel. The sensors and gauges were complicated, probably expensive, and made the belts more complex. Moore reported that the sensors worked and if the belts went into widespread production, these improvements could be utilized. However, with the twenty-one-second flight times and the high reliability of the belts to date, the more accurate fuel gauges were really not necessary. The various other warning devices they tried did not prove as effective, and Bell said it was going to stick with the one it had been using all along.[49]

To keep the rocket belt in the air longer, Moore tried coupling the rocket belt with a paraglider. He wondered if this addition might provide a way for the pilot to stay aloft longer, or, in the case of an emergency, to glide down and land after losing power.[50] A rocket belt pilot could not launch with a paraglider on his back though, so Moore experimented with an inflatable one developed by NASA. Of course, this design required the pilot to launch with the rocket belt as well as the uninflated paraglider and a tank of compressed nitrogen for inflating the glider.[51] The program was beset with problems from the start. Moore built inflatable paraglider wings, but they burst on inflation. Making them burst-proof made them heavier. Tests proved that the wings were unstable. They also tore.[52] The team managed to fly the paraglider with a man onboard by towing the device behind a car. Then the engineers discovered that the wings' material also ripped when landing in a field filled with weeds.[53] They decided it wasn't safe to test the paraglider over land. They moved their tests to a nearby lake. The engineers who had built a rocket belt with 100 percent dependability were looking at attaching it to a device they had no confidence in. This part of the program was soon scrapped.[54]

Another goal of Phase III testing investigated the difficulty of training operators to fly the rocket belt. How would new pilots be selected, and was there a way to streamline the training process? Since Moore had already trained Hal Graham, it was necessary to find someone with no experience and train him to fly the rocket belt. The new recruit this time was a nineteen-year-old named Peter Kedzierski.[55] Kedzierski had studied aviation mechanics in high school and was attending engineering classes at night school. Like Graham, he was athletic and coordinated,

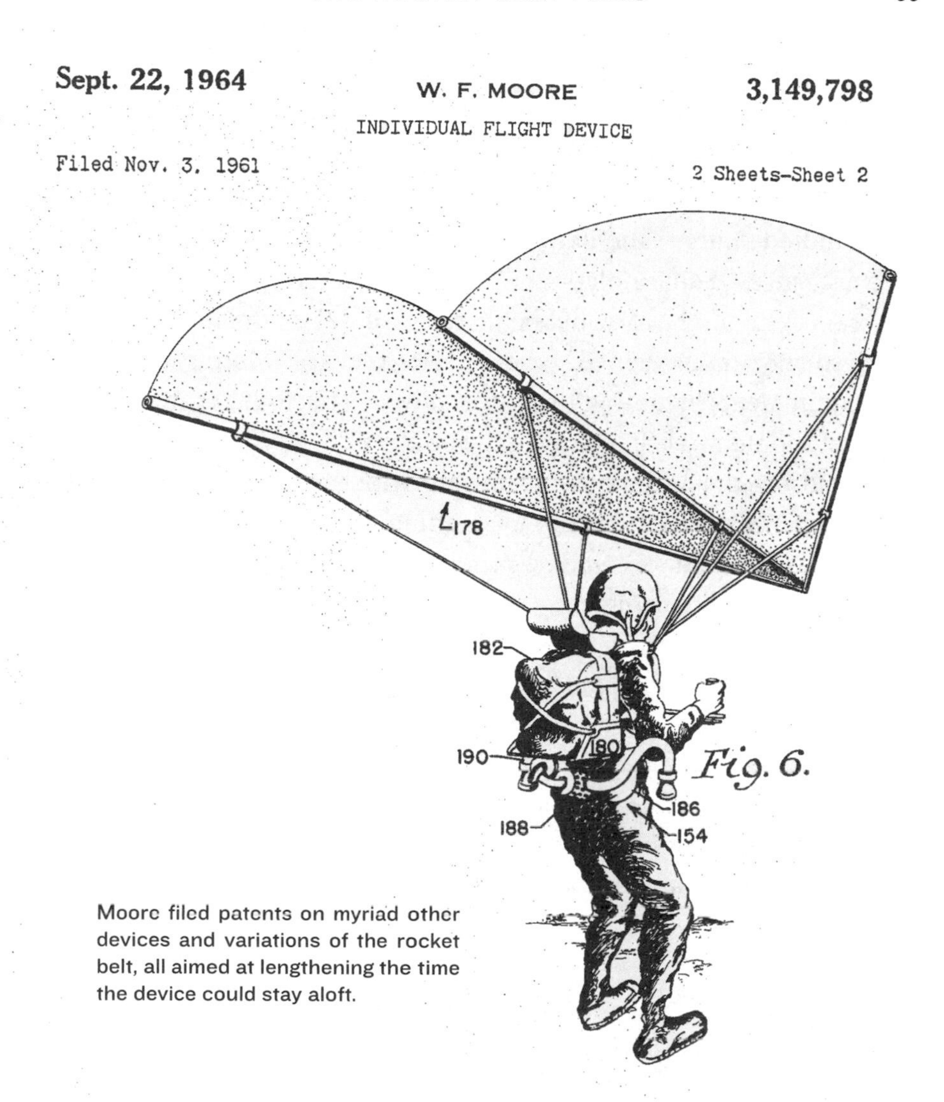

Moore filed patents on myriad other devices and variations of the rocket belt, all aimed at lengthening the time the device could stay aloft.

and Moore ran him through the same training process. Inside the huge hangar, they set up the tethers and let Kedzierski learn the controls of the rocket belt. As with the others, he started out shaky, with short hops and awkward landings. Moore duly recorded each attempt in a flight log and by Kedzierski's twenty-seventh attempt, he was ready for free flight. On March 2, 1962, Kedzierski became the second person to fly an untethered rocket belt when he made a short two-second hop and then a forty-foot flight where he touched his feet to the ground briefly at the halfway point.[56]

With this success, Moore informed the army that Bell was confident in its training program. After quizzing the engineers "most intimately associated" with the program, Moore listed the criteria for ideal rocket belt pilots. They would be eighteen to forty-five years old, weigh no more than two hundred pounds, and have a stable temperament with a "willingness to take chances." Some athletic ability was thought to be helpful and a couple years of high school would be nice. Beyond that, the most important requirement simply seemed to be a desire to learn. Moore summarized the requirements by noting that anyone who passed the army induction physical should meet the needs of the program.[57]

In closing out the report on the third phase of the testing, Moore noted that steps could be taken to lengthen the flight times of the rocket belt but not much more could be done with the simple hydrogen peroxide setup they had been using.

> A preliminary look at the use of jet engines for flying belts instead of fundamental rocket power indicates the greatest practical range achievement. The empty weight, however would be considerably greater and more complicated than the rocket-powered versions. This, however, would be more than offset by the jet engine's ability to use kerosene as a fuel and would lead to more widespread tactical use.[58]

Thiokol had suggested jet engines in its original report on individual lift devices. The problem was that no one built jet engines small enough for a man to carry on his back.

Bell developed the rocket belt with a focus on selling it to the military, yet Wendell Moore carefully protected his intellectual property claims with his invention. Moore was granted his patent for the rocket belt. Titled simply, "Propulsion Unit," patent number 3,021,095 was made official on February 13, 1962. In the next few years, Moore would apply for and receive more patents along with some of his coworkers at Bell for variations and improvements on the rocket belt.[59] Even though others were listed on the patents alongside Moore, there is no question that he was the driving creative force behind the rocket belt development. A proposal submitted by Bell to the US government in 1964 referred to Moore as "the inventor of the design evolved at Bell Aerosystems Company in 1957, and [he] did the original research and development which established the present configuration." [60]

In November 1962, *Popular Science* ran a short article on a device built for the air force by Chance-Vought, the company Charles Zimmerman worked for while developing his Flying Shoes. To help astronauts move in outer space, Chance-Vought had developed a unit worn on the back containing a nitrogen and hydrogen peroxide motor similar to the one in the Bell Rocket Belt. The device was called the Self-Maneuvering Unit (SMU). The unit was not designed for flight—the men wearing them would be in zero or near-zero gravity—but for maneuvering. The unit was large but it also contained breathing apparatus. Ten jet nozzles blasted the unit this way or that and would allow for astronauts to safely go on space walks and maybe even stroll on the surface of the moon. There, the SMU would help an astronaut by giving an extra boost as he walked or jumped in the lesser gravity of the moon. The article describing it was only a single page but the editors made history when they published it: they coined the term "jet pack."[61] Of course, this jet pack was not the same one that would commonly be known by the term. Still, from this point forward, people would increasingly speak of jet packs and use the term interchangeably for all manner of individual lift devices, whether powered by solid fuels, liquid-fuel rockets, or jet turbine engines.

7

THE BELL ROCKET MEN

Bell continued testing and building rocket belts and trained a team to show them off.[1] Hal Graham, however, stopped flying the rocket belt after an incident that is still the subject of some controversy. Reports on the rocket belt program insist that of the thousands of flights made by the rocket belt, no flights ever ended badly. Bell always reported a 100 percent success rate. Some critics would point to the accident that broke Wendell Moore's knee, but Bell would argue that was a tethered test flight. Still, there was another rocket belt accident. In 1962, Graham was traveling the country doing demonstration flights. At Cape Canaveral they wanted him to launch from a platform that was held in the air by a forklift. Graham was unaware that the platform was not sturdy and that no one had tested it to see what would happen when the rocket belt was blasted off it at full throttle. When he attempted to take off, the platform shifted and Graham lost his balance before he was completely under power.[2] He fell twenty-two feet to the ground and hit his head so hard his helmet broke. He was unconscious for half an hour—and would never fly the rocket belt again.[3]

The story was never mentioned when talking about the reliability of the belt. It would not be well known until more than forty years later when Graham told a gathering of rocket belt aficionados about it. In a widely read article written in 1963 by one of the highest-level managers of Bell, Graham's accident was glossed over. "Although never seriously injured during his tenure, he did have a few bad spills and collected many bruises."[4] Graham left Bell shortly after the crash. In all, he made eighty-three free flights with the rocket belt, many of them before large crowds and VIPs.[5] If Graham's flights had not gone as well, the program most likely would have floundered. In later years, fans of the rocket belt referred to him as "His Eminence."

Moore trained a few other men to fly the rocket belt. Joining Peter Kedzierski were Robert F. Courter Jr. and Gordon Yaeger (no relation to Chuck Yeager). Not everyone Bell hired made the grade. One pilot was trained but after a few harsh landings, Bell decided it had to let him go. Another was a former army flight instructor. He, too, made a bad landing in public and was fired. There were at least three rocket belt pilots who made it through the training phase of the program only to be let go because they committed bad performances in front of an audience.[6]

Bob Courter would become a major figure in the individual lift device field. Like some of the others at Bell, he had gone to school and studied engineering but had not graduated. In 1945, he trained as a pilot for the US Army and became a flight instructor. After his discharge, he moved to the Buffalo area and trained student pilots at Niagara Falls Aeronautical Corporation. In 1951, his instructing duties were interrupted by the Korean War. The New York Air National Guard activated him, sending him to fly several hundred hours of combat in a P-51 Mustang over Korea. By 1953, he was discharged as a first lieutenant. He was specifically hired to replace Hal Graham and would be in charge of the rocket belt pilots.[7] When he arrived at Bell, he was a combat veteran, certified pilot, and licensed to fly single and multiengine aircraft. He also held his instructor's rating. He first flew the rocket belt in July 1962.[8]

In 1964 a promoter named Clyde Baldschun contacted the marketing department at Bell. Baldschun booked events for state and county fairs, and his teenage son had read about the rocket belt.[9] Previously, Bell had turned down many requests to demonstrate the belt for the public because it was still testing the belt for the army and the cost was prohibitive. Baldschun assured Bell that he could book events for the flying belt at fairs and everyone would come out ahead financially. Bell's marketing department realized this might be a good way to keep the rocket belt in the public eye. Perhaps it would help with Bell's lobbying efforts for funding. The rocket belt team was soon at the 1962 Calgary Stampede. Bell charged the Stampede $25,000 to put on a series of flights over the period of a week, and Baldschun got a 20 percent cut for his efforts.[10]

Bell sent Courter out with the others to demonstrate the rocket belt far and wide, often to places Baldschun had booked. Courter and Kedzierski took turns wowing crowds in Chicago. Photos of the two accompanied a story in the *Chicago Daily News* quite typical of the rocket belt tour. "Visitors to the Chicago International Trade Fair are getting a look at the future through daily rocket flights by two performers. Several times daily the rocket men, powered by strapped-on Small Rocket Lift Devices (SRLD), soar 60 feet from the lakefront just east of McCormick Place."[11] It was a good year to see the rocket belt. The device was flown 396 times in public in 1964, including 170 flights at the world's fair. Bell estimated that two million people saw the belt fly.[12]

Bell continued pitching the rocket belt to the military, presumably because it knew that the government had deep enough pockets to fund a much broader research and development endeavor. In March 1963, *Science and Mechanics* ran an article expounding on the fantastic military capabilities of the rocket belt. The article, "America's aerial foot soldiers," was a mix of fact and fiction. Photos of the belt were interspersed with an artist's illustrations of the belts used in combat. Many of the images depicted the physically impossible. One showed a soldier flying over a battlefield while firing a bazooka. Another showed a soldier firing a gun at a jet—flying only a couple stories above a street between high rises! The rocket belt in each illustration was apparently controlling itself.[13]

The article described a recent flight by Peter Kedzierski. He had taken a ride up to a forty-five-foot altitude and covered 815 feet. At one point, he was said to be traveling sixty miles per hour. The article does not mention how short the flight lasted. Instead, the article focused on how a soldier could shoot at planes from "nine stories" up in the air, a height no rocket belt had ever flown. The final illustration showed American rocket belt pilots landing on a beach, but something ominous was on the horizon: The enemy was coming at them wearing rocket belts![14] The message was clear: if we don't start equipping our soldiers with these space-age devices, our enemies will start using them against us!

Kedzierski's speed and distance—sixty miles per hour and 815 feet—were records for the rocket belt at that time. A few months later, he and Courter flew at the Paris Air Show. This was the first time the belt had been demonstrated overseas.[15] When *Flight* announced their appearance, it gave the flight ranges of the belt as 815 feet, with a top speed of sixty miles per hour, and top altitude of sixty feet.[16] Bell scheduled fifteen flights at the show to be made by Courter and Kedzierski.

The trip to France involved some high drama. The fuel for the rocket belts was shipped to the docks at Le Havre where some dockworkers ignored the KEEP UPRIGHT warnings on the vented drums containing the hydrogen peroxide. As they rolled one of the drums along the wooden dock, it began leaking and caught fire. After a few panic-filled minutes, the fire was extinguished; the dockhands might not have even realized how close they came to blowing up the dock. Years later, Courter pointed out that the workers, if they had known enough about the fuel, could have simply rolled the leaking drum over the edge of the dock and into the ocean. The water would have neutralized the hydrogen peroxide.[17]

Writers from *Flight* in Paris were as impressed by the rocket belt as the writers who had seen it before them. Its first flight was short and low, covering only a thirty-yard distance out-and-back at a height of no more than five feet, but the belt's potential seemed clear to the audience. And while many people who saw the rocket belt elsewhere commented on how loud it was, the writers at the Paris Air Show said the belt caused "relatively little noise."[18] A photo in a later edition covering the events of the air show depicted one of the rocket belt pilots hopping "over the American space-exhibits dome."[19] The writer of the piece did not get a good look at the rocket belt in flight, though. The "rocketeer" launched so suddenly that he only caught a glimpse of it. The rocket belt's capabilities played into the disappearing act: the writer lost sight of the pilot "behind the crowd," a place where the rest of the aircraft at this show could never hide.[20]

In October 1964, *Popular Mechanics* ran one of its most expansive pieces on the rocket belt. Courter spoke with the writer and was even credited as cowriter of the piece called "I Fly the Man Rockets." Courter had now flown more than 250 times—the other Bell pilots had flown more than 550 times combined—and wanted to tell the readers what it was like to fly "like the birds." He admitted that the longest flights of the rocket belt were only twenty-one seconds, but then he quickly moved on to the control process. His right hand worked the throttle and his left hand held a control that turned the jetavators, which redirected the thrust to help him turn in mid-air. Then, he simply gunned the throttle and in a deafening roar, he was lifted into the air. It was as if "a giant had put a hand beneath my arms and another where I sit, jerking me off my feet. In an instant, I'm airborne. Next, I'm flying like the birds."[21]

The piece carried photos of Courter sixty feet in the air, flying over treetops, and images of him demonstrating the rocket belt at the 1964 world's fair in New York. Courter then went on to make some bold predictions, coupled with some secretive-sounding hints about things to come. "Man rockets—and their high-rocketing wearers—will be as common as helicopters in the decade to come. And, if anything, even more versatile." The rocket belts would be used by firefighters, surveyors, construction workers, military patrols, sportsmen, and fishermen, getting them to the hottest fishing spots. "They may even fly commuters over the bumper-to-bumper traffic."

Courter was not wrong in suggesting that people at Bell thought the belt might someday be used by everyday people. Wendell Moore's daughter later told a writer, "That was his dream—that it would become like a second car. That and to have the army pick it up. It was like a vehicle of the future for him."[22] Bob Roach, a project engineer for the rocket belt program, wrote an article in 1963 about the history of the Bell project up to that date. When describing the potential of the belt he pointed out, "Some see it as the answer for the traffic-jammed commuter." However, he suggested that more practical uses for the technology were in outer space or on the surface of the moon.[23]

The readers of *Popular Mechanics* should have noticed that many of the things Courter described were impossible with the limited twenty-one-second flight time, regardless of how inexpensive the devices became. Courter then dropped a bomb; he claimed Bell was on the verge of a huge breakthrough in rocket belt technology. "I base these predictions on a breakthrough, revealed here for the first time. A new, more efficient fuel promises to give Bell's man-hefting rockets 50 times their present range, up to perhaps 10 miles of free flight at 60 M.P.H. or better. Technically, the new fuel catapults man-rockets from the realm of a limited-range prototype to workaday reality."[24]

Courter did not couch his language or caution readers that the flying belt might not become widely available. On the contrary, he suggested they were right around the corner. "The day is closer than you may think." They might be pricy, but not prohibitively so. They are "apt to be Cadillac-priced." And they were easy to fly! There was no pilot's license required, and the FAA didn't even regulate them. "The learning is not much harder than mastering two-wheel bike riding. Most of my teammates soloed after 10 minutes of tethered practice."[25] Of course, ten minutes of tethered practice on a device that can only fly for twenty-one seconds at a time amounted to more than twenty-eight flights. Courter left the math for the reader to figure out.

What fuel was Courter talking about? It's hard to say. Bell never introduced a new fuel for its rocket belts although it would have been nice if they had found something as potent as Courter claimed. Hydrogen peroxide was dangerous and volatile stuff. Every second of burn time for the rocket belt consumed one and a half quarts of hydrogen peroxide; that small amount expanded into seventy thousand quarts of steam in a single second. Bill Suitor likened it to one milk carton of fuel expanding to ten school buses full of steam "faster than you can blink your eyes."[26] As one might imagine, anything packing that much of a punch is tricky to handle. Most materials commonly found in clothing will cause a reaction; the fuel would burn human skin on contact.[27]

Bell was always careful to only use the highest-grade nitrogen to push the fuel into the motor. Compressed air, such as that commonly found in SCUBA tanks, would contain too many impurities to be safely used.[28] Hydrogen peroxide is a strange beast. The kind Bell used came in thirty-gallon aluminum drums, enough for five to seven flights. The only thing that counteracts the chemical reaction of the fuel when it is ignited is water. The fuel handlers always kept buckets of water on hand to smother any hydrogen peroxide that got out. The aluminum fuel drums were designed to rupture rather than explode in the case of an unsafe buildup of gasses. In those cases, one pilot warned that you would still be in danger as the fuel spilled out of the broken drum: "Your shoes are one of the first victims; always be prepared with lots of water."[29] In the *Popular Science* article Courter described the revolutionary new fuel as best he could, considering that it was "still classified." He said it would allow the rocket belt to fly for ten minutes—almost thirty times longer than the current flights—and that flights could conceivably run into hours in the near future.[30] It's hard to imagine that Courter was simply making this all up. It also doesn't seem possible that he would be allowed to tell such a tall tale while still speaking on behalf of Bell, which he noted he was doing early in the piece. It can only be that Courter was referring to another advance Bell was working on, but it wasn't a new fuel. It was a new engine. Thiokol had described the possibility of a flying belt powered by a small jet engine in their 1959 report. Of course, in 1959, Thiokol said it could not find anyone who could build an engine small and powerful enough to launch the flying belt. By 1964, Bell had found someone: Dr. Sam Williams of Williams Research, in Walled Lake, Michigan.

8

THE SUD LUDION AND THE POGO

While Wendell Moore and his team at Bell were perfecting the rocket belt, other inventors worldwide were trying to develop their own individual lift devices. In 1960, a Frenchman named M. Caillette applied for a patent in France on an individual lift device that would come to be known as the Sud Ludion. Sud was a French aviation company, and Ludion is the French word for a Cartesian diver, a scientific demonstration where a small object is suspended in a liquid and then rises and falls based upon pressure applied to the container holding the liquid.[1] Presumably the Ludion that Caillette sought to patent would rise and fall just as easily as a Cartesian diver, at the control of its pilot. It took him a few years, but Caillette finally got the French army interested in his invention and they agreed to buy one from him if he built it.

The Ludion was a vertical take-off and landing (VTOL) device like the rocket belts of Bell, but it had one major difference. Instead of being powered by hydrogen peroxide, it used isopropyl nitrate. Isopropyl nitrate is a colorless liquid that is highly flammable, burns with an almost invisible flame, and was used as an engine starting fluid. It fell out of use because it is fairly dangerous and difficult to handle safely. Even so, it packs a punch, and Caillette thought it might get his Ludion off the ground. Isopropyl nitrate has a specific impulse one-and-one-half times that of hydrogen peroxide; that is, compared to an equivalent amount of hydrogen peroxide, the isopropyl nitrate produces one and a half times the power.[2]

French authorities gave Caillette a series of thresholds he had to meet with his device before they would consider it successful. It had to be able to carry a payload of sixty-six pounds along with the pilot, be able to travel over thirty miles per hour, cover several hundred meters, and be able to reach over 164 feet in altitude. Those requirements would have been prohibitive if applied to the Bell Rocket Belt, but Caillette thought he could pull it off with his Ludion. He agreed to the terms and then the army tacked on more requirements. The Ludion would need to be able to take off in a stiff crosswind and be able to survive a harsh landing, as if dropped from ten feet. The Ludion could no longer simply be a device strapped to a man's back. Caillette decided to build a small airframe around a seat, place the engine on the back of the unit, and provide it with a shock-absorbing skid.[3] At this point, the unit shared the VTOL characteristics of the Bell Rocket Belt and little else. Although the Ludion would be powered by a chemical reaction, the fuel was vastly different from what breathed life into the rocket belt. Still, the Ludion appeared quite similar in appearance to the Bell Rocket Belt—the pilot was simply seated instead of standing.

A version of the Ludion was built with air hoses that ran to its motor. For the tethered flights, the pilots flew using compressed air, much as Wendell Moore had tested his belt with the nitrogen rig. Tethered flights proved that the concept was viable: the Ludion flew. Caillette was asked to build a working model of the Ludion powered by the isopropyl nitrate. In February 1968, a Ludion made a tethered flight under its own power. The designers of the Ludion couldn't contain their excitement and their lack of patience caused them problems. Before they had even tested the device, they had brought it to the 1967 Paris Air Show and simply displayed it statically. Even though they hadn't gotten it off the ground yet, they had promised to demonstrate it in flight at the 1969 Paris Air Show. Someone even suggested they might fly two, in formation.[4] The desire to oversell experimental aircraft apparently was universal.

Sud was not secretive about the development of the Ludion: a detailed description of the vehicle appeared in *Flight* in 1968. "In the same category as the Bell rocket-belt, the Ludion is a rocket-powered VTOL vehicle to lift a single man and his equipment, plus a payload of about 66 lb., over 600 yards at a height of 500-650 ft." The entry described the Ludion's engine and noted that "prototype testing is in hand."[5] The problems the French team encountered were similar to those Moore had seen. The engine was loud—although not as loud as the rocket belt—and the amount of fuel the unit could carry was not enough to allow lengthy flights. Even so, when the Ludion achieved free flight in 1968, it could stay aloft for forty seconds. Some noted that this was twice the flight time of the famous Bell Rocket Belt. The increase was possible because the Ludion pilot was not carrying the weight of the fuel on his back; it was being supported by the frame of the device. Even so, the French military wondered what use they could find for a loud vehicle that could still only travel for under a minute.

The Sud Ludion was interesting in that it appears to have been the only foray into individual lift devices by anyone outside of the United States that gained any traction. In later years, inventors in other countries would work on rocket belts and jet packs, but in the 1960s the field was still dominated by the American inventors.

Meanwhile, over at Bell, Wendell Moore and his team had been working on variations of the rocket belt that paralleled the Ludion. Bell attached their rocket engine and tanks to a chair and to a small platform. The chair was simply a fiberglass office chair. In February 1964, the engineers brought it into the hangar where they had flown the rocket belt and attached it to the tethers. After determining that the chair was stable, more or less, they took it outside and Courter flew it. Courter had watched Moore assemble the chair and thought it looked perfectly flight-worthy. Not everyone agreed. One Bell executive bet Courter a quart of whiskey that the chair would tumble in flight and not be controllable. Courter won the bet, and his flights made for some great film footage, which has since found its way onto the Internet.[6] He mentioned that the chair was not as maneuverable as the rocket belt, but it could still be adequately controlled by the pilot.[7]

The platform flying device was nicknamed the "Pogo." From the side, it apparently reminded some of a man on a pogo stick. Bell also designed and flew a two-man Pogo that, as the name implied, carried two men aloft. It was made of the components of two rocket belts fixed onto a Pogo stand. The four fuel tanks and two nitrogen tanks fit between the two passengers.[8] These experiments with variations on the rocket belt began in 1966, and Bell also worked on having the devices controlled by thrust vectoring. After two years of developing ways to steer the Pogos in the air by angling the thrust nozzles, the engineers determined the units steered best kinesthetically. It was even true of the two-man Pogo.[9] Moore showed the devices to various government entities, hoping for financial support. NASA awarded Bell a $250,000 contract in 1968 to develop a lunar vehicle incorporating some of Moore's new designs.[10] During this time, Bell pitched quite a few ideas to NASA, pointing out all of the potential of the hydrogen peroxide rockets in space. If given development money, the rockets could be used to power a lunar lander or even to transport astronauts on the surface of the moon.[11]

One idea Bell pitched became known by the acronym LEAP, for Lunar Escape Astronaut Pogo. The Apollo program's lunar lander was equipped to take astronauts down to the surface from the orbiting command module and then back up again. Someone suggested the Pogo as a backup—to be used in an emergency. Could the Pogo be modified to travel from the moon back to a lunar command module orbiting the moon? The Pogo was never used by NASA for such a purpose but tests were conducted and some engineers crunched the numbers to see what modifications would be necessary. The problem with this notion was that the command module orbited the moon at an altitude of almost seventy miles. The Pogo, as it was configured by Bell, could not possibly move a man that far, even in the lesser gravity of the moon. Bell could have overhauled the unit for the lunar escape mission, but it doesn't look like any concrete steps were ever taken in that direction beyond mocking up some models and talking about the possibilities.

Decades later, a Bell engineer familiar with the rocket program at Bell and the company's various NASA projects said, "There were a lot of ideas floating around Bell during those times. I would not be surprised that such an idea would have been floated out as a possible application and they were trying to get some study funds to investigate the possibility more completely."[12] Still, there is a wide gulf between what was accomplished—the Pogo that could fly for around twenty seconds—and what might have been accomplished if they had set out to do something completely different.

Bell already had a relationship with NASA after working on reaction motors and rocket engines for the space program. One of its related inventions was the Zero-G belt, which Bell had developed for astronauts to move about in outer space. Bell anticipated the need for a spacewalk device and had begun development work on the project in 1959.[13] Bell noted in a report submitted on the device that it was related to another project: the flying belt for operating in "one-g." Operating with no gravity presented different problems. With no atmosphere to supply drag, the operator's body might spin or turn from the forces of the belt and continue to spin or turn until an opposite force was applied. Bell tested the Zero-G belt aboard the C-131 aircraft, which simulated zero-g conditions, in April 1961.[14] Forces required to move a man in space were much smaller than those required on Earth, but many of the lessons learned from testing the rocket belt were helpful.

The Zero-G belt was powered by compressed nitrogen and was much smaller than the rocket belt.[15] Eventually, though, NASA opted for the much larger Self-Maneuvering Unit to allow its astronauts to walk around in space. The SMU was the jet pack built by Chance-Vought, a competitor of Bell.[16]

Moore continued drawing variations and different configurations of the rocket belt, including some with jet engines in place of the rocket motors. He filed for a patent on the jet belt on November 8, 1966. He also drew up versions with engines on a simple Pogo stand and bolted to a chair. The patent for the new "Personnel Flying Device" was granted May 7, 1968, and noted that the primary difference between this and previous designs was a "rigid frame . . . ridden by the operator."[17] Moore pointed out that with the weight of the device sitting on the ground instead of the pilot's body, it could be constructed a bit heavier and still get off the ground. One version he suggested had space for two passengers and was powered by twin jet engines. It was clear that the variations and possibilities were endless. Although Moore filed patents for all these variants, Bell would only ever manufacture the jet belt.[18]

As the devices evolved, each improvement seemed to bring another complication. Building the unit as a stand that supported its own weight meant it was not as easy to control kinesthetically. A pilot seated in a chair had to lean farther to get the rocket-propelled chair to change direction. Making the units with bigger frames also meant the units were heavier. This required more fuel or bigger engines to maintain similar flight times. Still, Moore and the others kept plugging away at the puzzle. They believed they would hit the right combination of elements sooner or later.

9

WILLIAM P. SUITOR: ROCKET MAN

In 1964, Wendell Moore lived next door to a young man named William Suitor. Suitor was just nineteen and was thinking about joining the army. He had mowed Moore's lawn and done odd jobs for him, and Moore apparently thought well of him. Suitor later believed that Moore knew his father did not want him to enlist, considering the state of affairs in Vietnam. One day Moore asked him, "Hey kid, you want an exciting job?"[1] He told Suitor he might be able to get him a job at Bell Aerosystems. Suitor knew Moore was an engineer, but Suitor had no particular schooling or training for working in the aerospace industry. He was intrigued, however. On March 26, 1964, Suitor became an employee of Bell Aerosystems. "Talk about being a lucky bastard in the right place at the right time."[2]

Moore thought Suitor would make a good rocket belt pilot—after all, it didn't require any particular skills beyond those a man on the street might have. Bell had already finished the portion of the program where a nonpilot had been trained to fly the rocket belt. The Phase III report describing the training of Peter Kedzierski had been submitted to the army in April 1963.[3] Bell was now sending out its rocket belt pilots wherever Clyde Baldschun could find them a paying audience. And it turns out that Baldschun was having no problem finding them. Suitor would soon be seen by millions.

Suitor underwent tether training indoors, with cables and pulleys to keep him from hurting himself. He started out with low pressure so he could get used to the controls and then with increased fuel loads until he had full tanks and all the power of the rocket belt at his disposal. One day, the throttle handle broke and he lost control of the rocket belt. Discovering the perils of uncontrolled rocket belt flight first hand, Suitor shot around the hangar—still attached to tethers—"like a balloon that you inflate and just let go of."[4] Suitor was undaunted. He continued with his tether training and soon mastered the final training maneuver: a tethered flight from one end of the hangar to the other and back with a 180-degree turn at the far end. Then, Moore sent him outdoors without the lifeline.

"I took to it like a duck to water. My first free flight was pretty uneventful. Your mind is so loaded with stuff you don't really think about what is going on. You just react."[5] Once he was suitably trained, Suitor was officially offered a contract as a "flight demonstrator" to fly the Bell Rocket Belt in the state of New York and elsewhere.[6] His base pay was $147.50 per week, for which he agreed to fulfill any reasonable tasks Bell would give him in connection with rocket belt demonstrations and exhibitions. Those tasks might include, according to the contract, "radio broadcasting, television and personal appearances and other forms of entertainment, sales promotion and scientific demonstration as shall from time to time be assigned to the Flight Demonstrator by Bell."[7] The contract required Suitor to answer to two departments: the small rocket lift department and marketing.

Whenever Suitor traveled more than fifty miles from the Bell facilities, he would be paid $73.75 per week on top of his regular salary as an incentive, and he also was given a small per diem while traveling. Perhaps a little more sobering were two sections of his contract that covered the insurance Bell took out on him. Bell paid for a $50,000 life insurance policy on Suitor, to protect his interests, and also got his permission to take out "life, health, accident or other insurance" on Suitor to protect Bell's interests.[8]

Suitor's first professional flight in front of an audience took place at the 1964 California State Fair.[9] It was also Suitor's first trip outside the state of New York. The *Sacramento Bee* did a piece on Suitor, and as with many of the Bell Rocket Belt team appearances, local newspapers couldn't write enough about the men with the space age contraption.

> TEENAGER IS ONE OF ROCKET BELT FLYERS
>
> Whoosh! A man in a white suit with a helmet on his head, a system of tanks on his back and heavy duty insulated tennis shoes on his feet rises above the ground.
>
> As he speeds overhead, the groundlings grasp a momentary feeling of being in the middle of a runway at San Francisco Airport.
>
> The fellow descends rapidly to earth, hovers a few inches above ground and becomes one of the pedestrians.
>
> During the grandstand show each night at the California State Fair and Exhibition, William Suitor shares the rocket belt flight demonstration with Peter Kedzierski.
>
> Suitor, 19 years old and from Youngstown, NY, has been "flying" since March and has made between 40 and 60 flights.
>
> As for rocket belt flying, shrugged Suitor, "I like it . . . it's like a big fan lifting you. There's no physical strain . . . it's hard to believe though."[10]

One of these early flights was a bit more eventful for the young pilot. His short flight was set to take him from the fairgrounds racetrack, past the grandstand, and over to a stage where he would land. When the lights were dimmed, he took off and they shined a spotlight on him. Midair, with his twenty-one seconds of flight time rapidly evaporating, he had to guess where the stage was. As he flew by the spectators, he caught glimpses of the stage when flashbulbs fired. He angled for the stage and came in too low. He caught the edge of the stage and avoided injury, but the blast from the rocket belt scattered sheet music and sent musicians scrambling.[11]

In 1965 Suitor and the other rocket belt pilots made numerous trips, including one to France in March with Gordon Yaeger.[12] Bell had agreed to furnish rocket belt pilots for the stunts in the latest James Bond film, *Thunderball,* which was being filmed there. Sean Connery had already shot scenes wearing, but not flying, a rocket belt as he escaped from the roof of a chateau in the film's opening sequence. Suitor and Yaeger tossed a coin to see who would get to go first; Yaeger won and let Suitor fly first anyway. They took turns flying the belt as the cameras rolled and wound up flying three times each. They upset the director when they insisted on wearing helmets when flying: Connery had shot his scenes without a helmet. Needing the footage, the director re-shot Connery wearing one too. In the movie, the series of flights were spliced together for dramatic effect, even though the on-screen flight lasts exactly twenty-one seconds. The sound of the rocket belt was cut out by editors and replaced with a recording of a fire extinguisher.[13] Perhaps the most unrealistic aspect of the scene is how quickly Bond removes the rocket belt after the flight and how smooth and unwrinkled his nice suit is despite the rocket belt's harnesses cinched around him. Then again, he considered it part of his wardrobe; he remarked to his on-screen lady friend, "No well-dressed man should be without one." Suitor later stood in for Gilligan when the character flew a rocket belt on *Gilligan's Island*.[14]

The rocket belt team, including Suitor, was soon crisscrossing the United States and traveling abroad, flying the rocket belt in diverse locations. The team made hugely successful appearances at the 1964 New York world's fair where Bob Courter and others flew over the amphitheater.[15] As usual, the rocket belt brought the house down.[16] Elsewhere on the grounds of the fair, Chrysler was giving rides to visitors in a jet car. The Chrysler Turbine Car was powered by a turbine engine and Chrysler had hinted that the cars might be sold to the public. When a Bell Rocket Belt flew overhead or circled the huge globe on the fairgrounds, it seemed like the world of tomorrow was really happening.

Courter was the captain of the Bell Rocket Belt team and would eventually fly the rocket belt on four continents. He flew at the Paris Air Show, over Super Bowl I, in Sydney and Copenhagen, and on the television show *Lost in Space*.[17]

While the movie and television shows got the Bell Rocket Belt into the mainstream media, hundreds of thousands of people, perhaps millions, witnessed rocket belt flights at state and county fairs, and almost any other event where promoters wanted something flashy and loud to draw attention. In September 1968, Bell agreed to provide sixteen rocket belt demonstration flights as entertainment for the Adelaide Fair in Australia. Bell charged the fair $10,700 plus round-trip airfare for the three-man team from New York to Australia. The fair also agreed to pay hotel costs for the team and to provide a locked or guarded secure location to store the rocket belt and its support equipment. The contract included details of the rocket belt and its supporting crew. The rocket belt and its accessories weighed 540 pounds, for which the fair had to pay shipping. The promoters also paid to transport three drums of hydrogen peroxide and three cylinders of nitrogen, and to pay the customs fees, $1,500, for getting the materials into Australia. The drums and cylinders weighed 2,436 pounds.[18]

Bell was careful to ensure that the fair always called it the "Bell Rocket Belt," and if print ads were used, the full billing was "Textron's Bell Aerosystems Rocket Belt."[19] As previously noted, the military contract for the rocket belt had ended long before the team visited all of these countries and impressed millions with high-flying stunts. At this point, Bell was simply generating public relations for their rocket belt program.

In between appearances at fairs or shooting television shows and movies, Suitor was also involved in testing the belt at Bell, where engineers were still trying to find ways to make the belt more practical. At one point, someone wondered if a rocket belt operator could fire weapons while flying. Bell filled some lightbulbs with flour—it wasn't going to put live ammo on the pilots—and attached them to model rocket engines that could be ignited by the pilot while controlling the rocket belt. The pilot would launch with six of these bulb "rocket-bombs" on his own belt and hover near a target. Bell parked a Jeep for Suitor to fire at and in his twenty-one-second flight window, he would let loose with the full complement of six rocket-bombs. Although Suitor became quite adept at powdering the Jeep, Bell never took the tests any further.[20]

Suitor flew the rocket belt for Bell in forty-two states and a dozen countries, including New Zealand and South Africa. He flew at the British Grand Prix, where he "drag raced" against a Formula One car in July 1966.[21] On one occasion, Suitor managed a flight of 866 feet. A few years later he flew 1,200 feet and hit an altitude of 140 feet. He even hit eighty miles per hour once in 1966, but he swore he wasn't aiming to set a record. He was simply trying to get back to shore after flying out a bit too far over the water.[22] His last flight in the Bell Rocket Belt was at the New York State Fair in 1969. In the days preceding and including Labor Day, "Bell Aerosystems men" flew three times daily over an extravaganza that included the mundane—livestock shows and polkas in the entertainment center—to the unusual, such as "Miss Electra," the robotic homemaker of the future. "Is she real?" more than half a million fair attendees were asked.[23] All the while, Suitor flew twenty-one times in front of and above huge crowds. Victor Borge performed at the fair and Nelson Rockefeller, then governor of New York, sat on a curb and watched Suitor fly. When Rockefeller and Suitor attended a luncheon at the fair, Borge told the audience to welcome "Mr. Rockefeller and Mr. Rocketfella."[24]

After his last flight at the fair, Suitor returned to Bell. Shortly after, he heard that the rocket belt demonstration program was going to be cancelled. He assumed he would never fly another rocket belt again. He left Bell and went to work for the New York Power Authority. Others who had flown the rocket belt drifted off to find work, although some of them stayed at Bell. Gordon Yaeger had flown the belt more than seven hundred times since his first flight in 1963. He stuck around Bell until 1987, ending his time as a quality assurance engineer.[25] Bob Courter also stayed at Bell after the rocket belt team was grounded. He would go on to fly the next generation of individual lift device: the jet belt.

The de Lackner Aerocycle placed its user in a precarious position, above counter-rotating blades. Although it was successfully flown several times, the Aerocycle also suffered a few spectacular crashes and the design was eventually abandoned. *Photo from author's collection.*

The Hiller VZ-1 flying platform proved kinesthetic control was possible. *Photo courtesy of the Hiller Aviation Museum.*

The Hiller VZ-2 was larger and had controllable vanes for steering. The military found little use for the platforms. *Photo courtesy of the Hiller Aviation Museum.*

The Hiller VZ-1E, the last and largest of the Hiller flying platforms. The larger the platforms became, the less they could be controlled kinesthetically. *Photo courtesy of the Hiller Aviation Museum.*

According to the press photo, the Thiokol Jump Belt would propel a man up a mountain and allow him to run as fast as a racehorse. The Jump Belt did not live up to expectations and never made it past the testing stage. *Photo courtesy of Peter Gijsberts.*

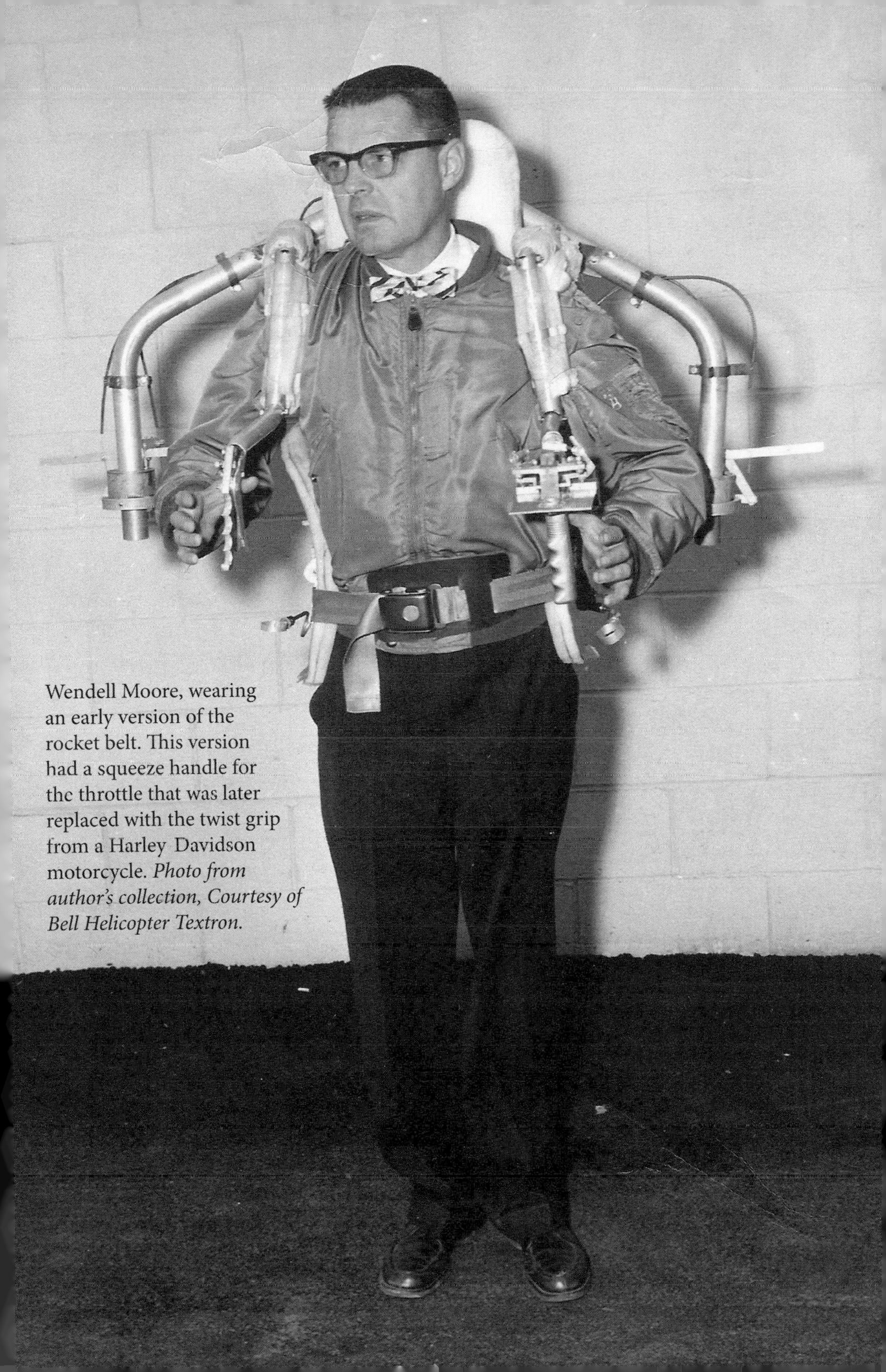

Wendell Moore, wearing an early version of the rocket belt. This version had a squeeze handle for thc throttle that was later replaced with the twist grip from a Harley Davidson motorcycle. *Photo from author's collection, Courtesy of Bell Helicopter Textron.*

Bob Courter in a press photo, wearing the rocket belt. Courter flew the rocket belt, the jet belt, the WASP, and the WASP II. *Photo from the author's collection, Courtesy of Bell Helicopter Textron.*

Bob Courter demonstrates the Jet Belt to the press in Fort Myer, Virginia. *AP wire photo, author's collection.*

Press photo of the Jet Belt on a stand. Because it was heavier than the rocket belt, a pilot could not comfortably support the weight of the unit with his body when it was fully fueled and not running. *Photo from author's collection.*

The Williams Aerial Systems Platform or WASP. This first version steered with thrust control and appeared less refined than the later WASP II. *Photo courtesy of Peter Gijsberts.*

10

THE JET FLYING BELT

Many of the people at the 1969 state fair in New York had heard of the Bell Rocket Belts before they saw Bill Suitor fly overhead, but most did not know that the belt they saw was already obsolete. A little over a year earlier, Bell had announced the next generation flying belt. This one was powered by a jet engine, and instead of being limited to twenty-one seconds of flight time, it would be able to stay aloft for ten minutes or more. At the moment Suitor was rocketing over the fair, the jet belt had already been free-flown by Bob Courter at Bell's facility in New York. But information about the jet belt and its capabilities wouldn't be widely publicized until November, when—who else?—*Popular Science* would run a cover feature on the device.

As early as 1964 it was becoming clear to engineer Wendell Moore and to Bell that the government would never be willing to buy the rocket belt in its current configuration. The twenty-one-second flight window was the most obvious shortfall of the device. Moore believed that the goal of a practical individual lift device was achievable, but the hydrogen peroxide–powered rocket's limited flight time was insurmountable.[1] Fuel tanks could not be made much larger without sacrificing maneuverability. And even if a pilot could carry twice as much fuel, how much more could he accomplish in forty-two seconds?

The answer lay in replacing the rocket motor with a small jet turbine engine. Thiokol had suggested the possibility in its initial report to the government on individual lift devices but had also determined that no one built a suitable engine at that time. In May 1964, Bell submitted an unsolicited proposal to the government pitching the rocket belt without a rocket;[2] Bell said it had found someone who could build a jet engine small and powerful enough to replace the hydrogen peroxide motor. In fact, two engineers at Bell, Wendell Moore and John K. Hulbert, had already applied for a patent on July 17, 1964, on what was essentially a rocket belt modified with a small turbojet engine in the place of the fuel tanks.[3]

Williams Research of Walled Lake, Michigan, had been founded by Sam B. Williams. He had spent some time developing a turboprop airplane for the navy, and then he worked with Chrysler at the inception of its automotive turbine program. Chrysler demonstrated its turbine car publicly in early 1954. Shortly after, Williams left Chrysler to start his own turbine manufacturing company, and he began designing and building smaller and smaller turbines.[4] Williams believed that turbines could be scaled down to be used in applications no one had dreamed of yet. In later years, experts noted that "without his invention of the small gas turbine engine, the cruise missile and affordable business jet aircraft would not have been possible."[5]

Sam Williams had a mechanical engineering degree from Purdue University and graduate training in the fields of thermodynamics and aeromechanics. At the time of the jet belt proposal, he had been in engineering management for twenty-one years, including the dozen with Chrysler working on turboprop engines and turbine-powered cars. Williams had developed five different designs of turbine engines, ranging from twenty-five to six hundred horsepower. He also held many patents on his designs and would soon be granted a patent for the turbine engine that Bell was now proposing for the jet belt.[6] He later said, "the way to progress further in this business was to step out and develop some new engines that would bring to the small engine field the same kinds of performance, thrust-to-weight ratio, and so on that were being achieved in big engines."[7] Williams had even successfully placed small turbines on the rotor tips of a helicopter.[8] Williams Research may have been small, but Sam Williams made a point to hire the brightest and most experienced engineers he could find. The team working for him on the jet belt turbine

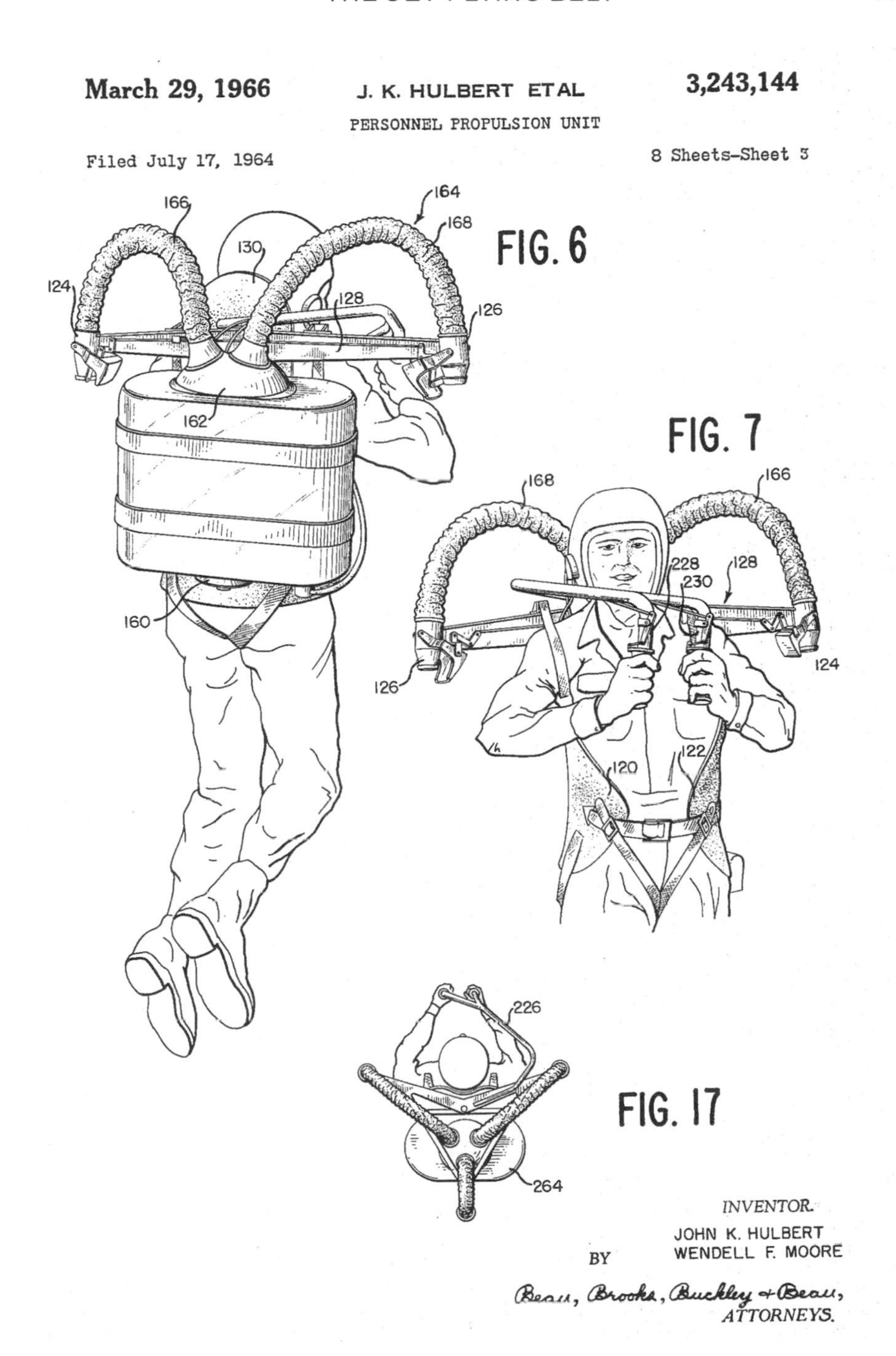

The patent for the jet belt built largely upon the already patented rocket belt. Bell patented the belt, and Williams was granted a patent on the engine.

included three graduates of MIT and engineers who had worked at Lockheed, Orenda, Continental, Allison, Chrysler, and Ford.

Bell's proposal claimed the engineering design work had been done. All that was needed was funding, not for the belt portion of the device, but to manufacture the engine. Wendell Moore proposed a simple swap—more or less—of the turbojet engine into the place of the hydrogen peroxide motor on the rocket belt. According to Moore, the design would be "easily controllable," and have a range of ten miles.[9] Moore also predicted a flight ceiling above ten thousand feet and a flight time exceeding ten minutes.[10] If he was right, the jet engine would be extending the flight time of the belt by almost thirty times.

Moore realized that the effect of an engine failure on the pilot of the jet belt needed to be addressed. None of the individual lift devices could glide or land safely if they lost power. Anticipating the question in the jet belt proposal, Moore drew comparisons between it and other devices the army used. For example, soldiers had been parachuting out of airplanes for years now. Between 1955 and 1956, there had been 688,211 documented parachute jumps by GIs with eleven fatalities reported. That was only one death per 62,500 jumps. Crunching numbers on how many hours GIs spent flying in fixed wing aircraft and helicopters resulted in one death every 32,350 flights, assuming each flight averaged an hour in length. Moore suggested that since the jet belt would fly for no longer than an hour, if Bell could limit catastrophic failures to only one per thirty thousand to sixty-five thousand flights, the jet belt would be safer than these other modes of transport. Moore was confident they could reach those numbers.[11]

The heart of the jet flying belt would be a small turbofan engine Williams had recently designed and was in the process of patenting.[12] Moore and Williams promised the army that Williams Research could manufacture an engine that weighed only sixty pounds and still develop 425 pounds of thrust. The engine would actually be run at a lower output, around the 405 pound range, to put less stress on it. Even so, this amount of lifting power was substantially greater than the three hundred pounds or so that the hydrogen peroxide–powered rocket belt could manage for only twenty-one seconds. And the new device would not need exotic or dangerous fuels to run: the jet engine could run on a fuel as common as JP-4 jet fuel, which is basically a mixture of kerosene and gasoline.[13]

Without commenting on how unusual such a setup might appear, Moore's proposal described how a jet engine would be strapped to a pilot's back as if it were nothing more than a rucksack for a long march.

> The propulsion system consists of a normally rated 425-pound thrust turbojet engine, mounted vertically on the back of the fiberglass corset. The engine is installed with the air inlet down and the outlet upward. To the outlet is attached a bifurcated controllable flexible nozzle assembly discharging downward on each side of the operator's shoulders, just as with the rocket belt's.[14]

The jet belt would weigh a little more than the rocket belt, however. The rocket belt had weighed 63 pounds empty and 110 pounds fully fueled. The jet belt proposal anticipated a weight of 90 pounds without fuel, and 123 pounds after filling the 5.35 gallon fuel tank. Suspecting that the army might still wonder what good a flying belt could do regardless of the flight endurance, Moore suggested that the new jet belt would have room to spare for twenty-five pounds of payload and still meet all the other promises made about flight time. He also said that the jet belt could fly a hundred miles an hour. In case someone doubted that, he noted that "our existing rocket belts have been flown at speeds up to 65 miles per hour with no noticeable change in stability or control."[15]

The proposed jet belt was pitched to do many of the tasks that had been identified as likely jobs for the rocket belt.

> The purpose of the Individual Lift Device or Jet Flying Belt is to provide a substantial improvement in individual soldier mobility for a variety of select mission applications such as (1) Observation (2) Reconnaissance (3) Forward observation and liaison (4) Overcoming natural and man made obstacles (5) Clandestine operations and (6) Delivery of personnel and parts for in-place maintenance.[16]

Meanwhile, Williams submitted a patent application for the new turbine engine on September 22, 1965.[17] The army hadn't officially asked for the jet flying belt proposal, but it showed the proposal to the Advanced Research Projects Agency—known as ARPA at this time, and sometimes as DARPA when associated with the Defense Department—an agency formed in 1958 in a reaction to *Sputnik* and fears within the government that other countries might be gaining a technological edge over us. ARPA was not affiliated with a single branch of the military and was solely interested in developing cutting edge technologies that might be necessary in keeping ahead of our international rivals. It would, in theory, remove some of the politics from the process of developing projects for the government, even if they might wind up in military use.[18]

The contract to build the jet belt was granted to Bell on December 30, 1965.[19] The funding came from ARPA, and the Army Aviation Materiel Command was asked to oversee the project.[20] Bell, working with Williams Research, would develop and demonstrate the jet flying belt.[21] The contract specification required that Bell deliver one jet flying belt, a spare engine, the necessary "residual hardware and data," as well as demonstrate a capability to fly the belt ten miles with a speed approaching sixty miles per hour.[22] While the jet belt would resemble a rocket belt in many respects, the device would have to be reworked from the ground up with an eye toward the unique power plant. The controls would appear to be the same from the pilot's perspective but that was where the similarity ended.[23]

The contract showed Bell as the primary contracted party, with Williams Research as subcontractor, to furnish the engine.[24] The jet belt patent, number 3,243,144, was awarded to Wendell Moore and John Hulbert on March 29, 1966.[25] Still, the key to the jet belt was the spectacular Williams WR2-2 engine that Sam Williams had been developing for almost ten years. Williams had initially developed the small turbine for the military to power a surveillance drone—a practically unheard-of technology at the time. Three million dollars had been spent in creating "the world's smallest turbojet engine." Bell and Williams Research proposed making modifications to the drone engine so that it would power the jet belt. To show how small the WR2-2 engine was, the proposal included a full-page photo of a young woman holding the engine in one hand and smiling, as if it were a simple household appliance.[26]

The proposal explained, and showed in photographs, how Williams had successfully put turbine engines into a boat, a Jeep, on the tips of helicopter rotors, and as an auxiliary power unit for an aircraft. Bell assured the army that the jet belt engine would not require Bell or Williams to create an engine from scratch; costly procedures would be avoided by "adding conservatively designed elements" to the WR2-2.[27] To power the jet belt, the WR2-2 would be increased in size a tiny bit; even so, the new engine would only be twenty-two inches long, and twelve inches in diameter.[28] The slightly modified engine would be renamed the WR-19. The engine components were made of titanium and stainless steel and included counter-rotating spools to minimize the torsional effects the spinning parts might otherwise exert on the jet belt.[29] And, while those counter-rotating spools kept the engine from wanting to twist as it ran, they gave the device more stability than the rocket belt by adding gyroscopic influences on the unit.[30]

The process for building the jet engine was a little more involved than what had preceded the development of the rocket belt motor. ARPA and the army followed the progress at Williams through regular progress reports and went so far as to conduct site visits. By November 1966, the first phase of the development was completed. The jet belt design had been finalized and it was just a matter of putting it all together. On paper, Bell reported that the jet belt would weigh ninety-nine pounds without fuel and would be able to fly the required distance of ten miles and hit speeds of one hundred miles per hour.[31]

Now it was a matter of assembling the parts of the jet belt, building and testing the engine, and then installing for Phase III, the flight testing. Williams had been testing the engine in a test cell in Walled Lake. At the time of its progress report, the engine had been run for a total of seven hours and up to 52,500 rpm, a little over 95 percent of its rated ability. Williams engineers had found a few minor issues but nothing that couldn't be resolved quickly. The engineers stated in the report: "Major mechanical problems have been nonexistent to date. Data and testing to date indicate the engine rated thrust and specific fuel consumption appear achievable in the present configuration."[32]

Interest in the jet belt was high, and not only because of the individual lift device application. The power plant attracted quite a bit of attention on its own. On September 26, 1967, a contingent from various government agencies visited Williams to watch a demonstration of the engine. Among the group were representatives from the Australian and British embassies and the Canadian Army.[33] Bell was hoping to be able to market the jet belt to our allies; some of the observers were thinking of other uses for the little jet engine. Perhaps it could power drones or cruise missiles.[34]

Bell announced the new project to the public and gave the press a briefing. "JET FLYING BELT IS DEVISED TO CARRY MAN FOR MILES," read a headline in the *New York Times*, accompanied by a picture of a man wearing the jet belt. Bell did not demonstrate it since the company had not flown it yet, but it showed the press films of the rocket belt flying and simply told them to imagine it staying aloft for a lot longer. While the public had seen flying belts previously, those who had followed the developments would have noticed how different the jet belt looked. Instead of small tubes coming out from behind the operator's head, the jet belt had huge exhausts that flanked the pilot and pointed to the ground. Large clear fuel tanks were also visible on the sides of the engine.

Robert May, the project's manager at Bell, explained to the press that while the rocket belt could only fly for twenty seconds, Bell was confident that the jet belt would be able to stay aloft for "many minutes" and be able to fly several miles before refueling. Bell also stressed to the press how safe the flying belts had been so far, having completed more than three thousand flights "without injury or accident."[35] Bell didn't mention Wendell Moore's fractured kneecap or Hal Graham's crash landing at Cape Canaveral. Perhaps indicating his disappointment in not getting to see the belt fly, the reporter closed his article by noting, "No one arrived by flying belt."[36]

Bell was being very careful with the development of the jet belt. They built a ⅝-scale development rig to test its theoretical flight characteristics before they tried to fly it. This process was partly a reflection of how expensive the jet engine was compared to the rocket motor. The components of the rocket belt had been largely off-the-shelf and inexpensive; the jet engine for the belt was extremely expensive, and Bell could not afford to have it damaged in a test flying accident. Meanwhile, Williams tested the engine at its facility in Michigan. Compared to the rocket belt, the jet belt literally would have much more riding on it. The rocket belt pilot had not been able to fly long enough to ever have to worry about durability issues on long flights. If the jet belt was capable of staying aloft for ten or twenty minutes, the pilot had to be assured that the engine was not likely to fail. Engine failure probably would be catastrophic, so Williams tested the engine thoroughly. In the summer of 1968, it finished the flight rating tests and sent the engine to Bell.[37] Wendell Moore placed the engine in the belt and began testing.

Sam Williams was always coming up with new ideas and innovations. By the time he retired he had more than seventy patents in his name, most related to jet engines. While working through the issues of the jet belt, he had also wondered about the steering controls. While the jet belt was configured with the same two-handed controls of the rocket belt, Williams wondered if there was a way to free up the hands of the pilot. So he designed a system that allowed the pilot to control the jet belt's flight with his head movements. When the pilot tilted his head forward, the exhaust of the belt would be vectored to move forward. If he tilted his head back, the thrust would be vectored to slow forward motion. Twisting of the head would result in yaw control, in the direction of the head turn. The system was never used on the jet belt, but the patent was filed in 1965 and granted in 1969.[38]

Another interesting issue arose with the jet belt. The rocket belt had not needed a starting device; the engine fired the moment the valve was opened and the nitrogen forced the hydrogen peroxide fuel over the catalyst bed. A jet engine takes a little more to get running. Something had to get the blades spinning fast enough for the jet process to start working. An elaborate starter motor would not work because of its weight. The easiest solution was a solid propellant cartridge. The cartridge contained solid fuel that generated hot gases; the engine was designed in a way that allowed the hot gases from the cartridge to flow into the turbine and push the fan blades to get them moving. This setup meant that the engine could not be restarted in flight. On the plus side, the system was simple and lightweight.[39]

In 1968, Bell hired an experienced military pilot named John Spencer to be the project pilot for an experimental aircraft called the X-22, a vertical take-off and landing aircraft powered by four huge ducted fans. Spencer had flown helicopters in the air force and British military and had gone to test pilot schools both in America and England. He had flown over a hundred different types of aircraft by the time he joined Bell, and before he left, he'd fly the rocket belt a hundred or so times as well. Spencer became the lead test pilot at Bell with the rocket and jet belts under his direction.

At the same time, Bell was also working on a NASA project for a vehicle to operate on the surface of the moon. Some people within the company began to worry what might happen if one of the rocket belt demonstrations went awry. To date, the flights had gone quite well, but one public crash could inflict incalculable harm on Bell's reputation. Bell decided to ground the rocket belt flying team. "Bell did not want to run the risk of having an incident at a public demonstration that would jeopardize the whole NASA concept. That's the reason the whole demonstration team was broken up," Spencer explained.[40]

Bell completed the NASA development contract but did not obtain the contract to build a working model. In the end it wasn't engineering shortfalls that kept Bell from sending its rocket devices to the moon; it was practicality. The lunar rover that won the design contest could only travel so far—if it broke down, the astronauts could walk back to the lunar lander. If the astronauts were rocketing about on the lunar surface, they could conceivably get too far to walk back to their capsule. NASA decided to go with the more pedestrian car-like lunar rover and forgo rocket technology for personal mobility on the moon's surface.[41]

Moore and others within Bell began to wonder if there might be something they could do to increase the odds of selling the products they were developing. They knew more was required than simply building a working jet belt; they had to figure out how to get someone to buy it from them. Like the rocket belt, the jet belt contract called for engineers to design, build, and fly the device. What would happen then? What if the military passed on the invention as it had with the rocket belt? To head off that concern, Bell did some market research by sending out a combination report-questionnaire. The report's introduction stated that Bell was "conducting a study to determine the characteristics most desired in a military Light Mobility System."[42] The report was apparently sent to decision makers in the military who were asked to read a summary of the flying belts to date, including a section on the expected performance of the jet belt, and to then tell Bell what they would like to see done with the technology. Bell included a self-addressed stamped envelope for each decision maker's convenience.[43]

The booklet must have been put together quickly. It included photos of the jet belt, assembled but not flying. For flying shots, Bell used photos of the rocket belt. To illustrate what the jet belt would be capable of doing, such as "reconnaissance or Hit-and-Run Missions," Bell included drawings of soldiers wearing jet packs, but the jet packs in the drawings did not look at all like the actual jet pack. The engine in the illustrations was much larger than the one Williams developed for Bell and it was shaped differently. The illustrations in the report suggested that the artists were not given access to photos of the jet belt as it was being developed and were told to simply imagine what the belts would look like.[44]

Bell outlined all of the various configurations it could build, if need be. After explaining that the company had developed and flown a rocket-powered chair, a one-man Pogo, and a two-man Pogo, the company explained that it could just as easily configure those devices with jet engines. The possibilities were endless. A two-man, jet-powered Pogo could be used to rescue downed pilots or bring command personnel to the front so they could get an aerial view of a battlefield. A one-man Pogo could be used for security and patrols. A jet belt operator could string wire or lay a smokescreen. It was clear that Bell thought the device had many potential applications. It was just a matter of whether the military would agree, and what configurations they would prefer.[45]

In July 1968, *National Geographic* asked Bell if the magazine could send someone out to photograph Wendell Moore and a rocket belt pilot in flight. One of their most revered photographers, a man named Dean Conger, went to New York for the portrait. Conger, Suitor, and Moore

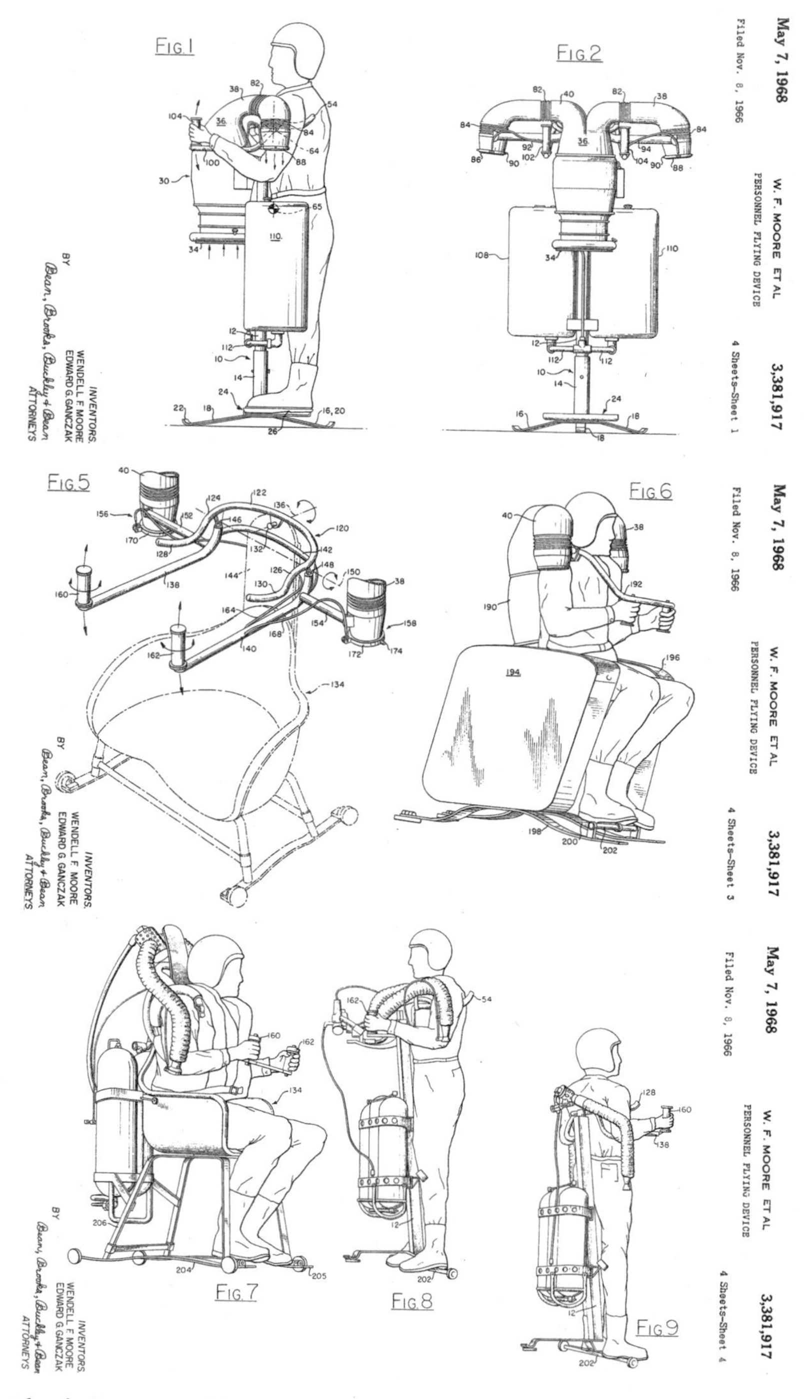

Patent for the jet-powered Pogo and other variations.

went back to the Niagara Falls airport where Graham had made the first untethered rocket belt flight seven years earlier. Conger snapped a photo of Moore in profile with Suitor hovering in the background. *National Geographic* planned on using it an upcoming issue.[46]

On April 7, 1969, Courter made the first untethered flight with the jet belt at Niagara Falls, New York.[47] At Bell, Courter had been given the title of Chief of Small Lift Devices Flight Test and Operations, and he wrote a firsthand account of his jet belt experiences for *Popular Science* in November 1969. As with the previous article he contributed to *Popular Mechanics*, this one was a mixture of science and public relations. Within a few sentences, he simply cut and pasted in an entire press release for Bell.

> The new mobility system will enable men to fly over rivers, cliffs and canyons, or such man-made obstacles as mine fields, barbed wire, buildings, and highway traffic. The scope of military and civilian applications is limited only by the imagination and ingenuity of potential users.
>
> Reconnaissance, artillery spotting, counterguerilla warfare, ship-to-shore operations, and perimeter defense are military operations considered within the Jet Belt's capability. Civil applications include traffic and riot control, fire-fighting, rescue operations, rooftop and high building patrols.

In case the reader wondered if the jet flying belt would ever be available to the man on the street, Courter believed Bell would be happy to oblige. "Conceivably, the Jet Belt could even become the commuter's vehicle of tomorrow. Maybe someday your 'second car' will be a flying belt garaged in the hall closet."[48] Courter noted for the readers that he had logged more than a thousand flights in the rocket belt before upgrading to the jet belt, and the jet belt had "vastly improved capabilities." Instead of flying time in the seconds and flight distances in yards, the jet belt measured its flights in minutes and its distances in miles. It is noteworthy that this article appeared soon after many people would have seen the rocket belt at their local state fair. Just a few months earlier half a million people had attended the New York State Fair, where Suitor had flown the rocket belt twenty-one times in a week. Events like that had been running nonstop since the belt was first proven viable. Now, Bell was announcing the next phase.

Interestingly, Courter noted that the jet belt was easier to fly than the rocket belt. The rocket belt pilot needed a larger ground crew to handle the sensitive hydrogen peroxide fuel. The jet belt's kerosene was relatively easy to handle and required less ground support. He then walked the readers through the process. He stepped into the "corset" and strapped himself in. The jet belt, when fully fueled, was a bit heavier than the rocket belt at take-off, so the unit sat on a small retractable stand that took much of the stationary weight off the pilot's back.[49]

Courter explained the trick to getting the turbine engine started with the solid fuel cartridge. From that point, Courter would give the engine enough throttle to maintain "ground idle." This was the engine speed that would allow the engine to keep running but not lift off the ground. He would check over his gauges to make sure everything was operating properly and then give the throttle a little twist, moving up to "flight idle." Courter noted that even at this low speed, the entire unit became noticeably lighter. Idling, he could move around on the ground with most of the weight being carried by the thrust of the engine. He would then retract the stand and give the belt more throttle. Then "I lift off, rising like an elevator."[50]

After first lifting off, while still within a few feet of the ground, Courter noted that the exhaust blasting down from his jet pack ricocheted back up at him. This meant each flight started out a little bit rough. As a result, he usually wanted to get up and away from the ground fairly quickly. From there, the unit operated like the rocket belt, but without the hectic concern that the pilot had to look for a place to land almost immediately. Speed and altitude were largely controlled by the right hand grip, which was twisted for more or less engine power. The left hand grip allowed him to rotate on a vertical axis. He could move up and down in the sky, go forward, backward, or side to side. He could also stop, hover, and rotate 360 degrees. He noted that while being "exempt from gravity," he felt "a sense of freedom," much like when he was scuba diving.[51] Although the fuel capacity of the unit was a bit larger, Courter usually operated it with the same total weight as the rocket belt. This meant flying it with tanks that were not completely full. Even so, flights of ten to twelve minutes were common with the jet belt.[52] Speeds of sixty miles per hour were also demonstrated.[53]

Unlike the rocket belts, the jet belt was only flown by Courter.[54] This did not sit well with the other rocket belt pilots. Years later, Suitor told a writer he wasn't even invited to watch Courter fly the jet belt. He found out about the first flight of the jet belt when he heard the jet engine being fired up while he was running an errand at Bell. He was still employed by Bell but no one had bothered to tell him that the jet belt was about to be flown.

> I looked around the corner, and there was the God-damned Jet Belt hovering. It was like somebody kicked me right in the stomach, you know? I was supposed to be part of this, we were all in this together, and here they were flying the son of a bitch and they didn't have the common decency to invite us to come watch. I could have sat down and cried—it really, really hurt. There were some traffic cones on the grass, and Courter was about six feet in the air. I stood and watched him for a few seconds and then left, in disgust really. I never saw it after that.[55]

The decision to have one pilot was simply a matter of cost, according to people who worked on the program. To train another pilot would have required time and money, and it would have put more wear and tear on the jet belt. Even though the jet belt was similar to the rocket belt, it was different enough to require more training. Bell figured the risk wasn't worth it. The only downside would be that if Courter got sick or injured, Bell would not have anyone available to fly the belt. Luckily, Courter never missed a flight.[56]

Wendell Moore lived to see the final version of his flying belt: the jet belt flew on April 7, 1969. On May 29, Wendell Moore died of a heart attack at the age of fifty-one.[57] He had never free-flown one of the belts; the closest he came was the tethered flights, which ended with the fall and the broken knee cap. Not being able to fly his rocket or jet belts had been a major letdown for him. Bob Roach said, "I think that was his greatest disappointment, that he never got to make a free flight."[58] His fellow engineers marveled at what Moore had accomplished at Bell. "What Wendell did was remarkable," Roach said more than forty years later.[59]

The military told Bell that while it was happy with the fulfillment of the jet belt contract, it did not see a practical use for the belt.[60] Operating a jet engine in the battlefield required a trained crew and supporting equipment. If the device was to compete with a Jeep—which many evaluators had compared it to—it couldn't win in a head-to-head competition for ease of use. A GI can jump in a Jeep on a moment's notice and drive off with no special training or equipment. The support structure for a Jeep was likewise minimal. The two didn't compare well. The jet belt "could never be a rough-and-ready kind of a thing a guy could throw in the trunk of his car and drive it to the beach or drive it someplace else and then put it on and go fly away."[61]

While many of the Bell engineers on the program thought the flying belts held potential, skeptics worried about the shift from rockets to turbines. Turbine engine technology was the specialty of Williams, and the Bell engineers thought they could do nothing more with the program from this point forward. They offered to sell the jet belt to Williams, along with the licenses for the patents, so that Williams could continue its research. Williams bought the license for the jet flying belt from Bell on January 23, 1970, for a reported $5,600.[62] Sam Williams had a fascination with individual lift devices and didn't want to see the program end. He wanted to try a few experiments of his own. Some might wonder why Williams didn't simply start manufacturing and selling jet belts in light of the successful flights Courter had demonstrated with it. Among other factors, one was probably cost. One source suggested that the cost of the engine in the jet pack was in the range of $85,000—in 1980 dollars.[63]

Flight reported the transaction as an agreement by Williams to sell the jet belts for Bell: "In addition Williams Research signed a five-year agreement with Bell Aerospace in January 1970 for WR to manufacture, use and sell the Bell Jet Flying Belt powered by a WR turbofan."[64] It is not clear where *Flight* got that information; there is no evidence that anyone ever agreed to have the jet belts manufactured for sale.

It was around this time that Suitor left Bell and went to work for the New York Power Authority. He had flown the rocket belt more than twelve hundred times.[65] Courter followed the jet belt to Williams Research and became a test pilot there. John Spencer also left Bell but stayed in touch with Bob Courter, who asked him if he was willing to step in for him as a backup pilot on the jet belt, in the event something were to happen and he couldn't fly it on some important occasion. Even though Spencer had never flown the jet belt he was confident he could learn it relatively quickly if needed. Courter had maintained not only his pilot's license but also his flight instructor certification while working at Bell and Williams.[66] If the time ever came to train others, Courter would be ready. The opportunity never presented itself, however.[67]

Williams Research continued experimenting with the jet belt with Bob Courter at the controls. Along with the longer flight time capability, the device also had the ability to fly higher than the rocket belt. Some people had revisited the obvious question as to what might happen to the pilot if the engine stalled in flight. Williams installed a ballistic parachute on the unit so that in the case of an emergency, Courter could trigger an explosive device to deploy a parachute very quickly. If he was higher than seventy-five feet or so, the parachute would lower him to the ground. Under seventy-five feet however, the device wouldn't have time to deploy and the risk to the pilot remained.

Williams Research had been developing other products while working on the jet belt, including other applications for the jet belt engine. *Flight* noted in 1971 how useful the little turbine engines would be in other devices, such as "drones and pilotless aircraft with short-life or expendable engines."[68] They were right: the WR-19 engine that had powered the jet belt turned out to be just the thing the military needed to power the Air Launch cruise missiles and the Tomahawk cruise missile. The engine was modified somewhat and renamed the F-107. When powering the missiles, it put out more than six hundred pounds of thrust. The F-107 was a bit more sophisticated than the WR-19 and a little bit larger in some respects. It would not fit into a jet belt; to fly now, it would need a bigger frame.

Bell and Williams were not the only ones thinking about jet belts. No one else built any around this time, but at least one inventor thought he had something to say on the topic. In 1977, Thomas Moore, the man who had designed the jet vest in Alabama, filed a patent for what he called a "Man Transport Device." Strangely, his invention was illustrated as a backpack with two small jet engines, one on each side of the pilot. The patent description said that the configuration of his device would allow for greater range and more simplicity than previous designs. He did not explain who would make the little jet engines he drew into his illustrations, or how they would be fueled. His drawings did not show fuel tanks. Instead, it looked more like he was trying to drum up interest so he could get some government funding. The patent began with a Dedicatory Clause offering his invention to the government "without the payment to me of any royalties thereon." The US Patent Office does not require inventors to actually build their inventions and many patents are granted on devices that have never existed beyond the paper upon which they were drawn. Some patents are even granted on devices that would never work. Moore's "Man transport device" patent was granted on August 9, 1977.[69]

11

THE WASP

Sam Williams was still interested in the field of individual lift devices, particularly if it would provide a market for his small turbine engines. The people who had seen the jet belt fly knew that it was revolutionary, especially in light of the huge leap forward it represented over the rocket belt. But the jet belt still had limitations: the weight of the fuel and the weight of the device at take-off made it unwieldy for all but the strongest of pilots. It had even gained a few pounds since Williams acquired it because of the additional ballistic parachute. It certainly wasn't a handy thing to heft while on the ground. The next step in the evolutionary process seemed obvious: a flying platform. The flying platforms built by Hiller back in the 1950s had been powered by heavy piston engines spinning propellers. What if the platform was powered by an engine similar to the one in the jet belt or the variant being used in the cruise missiles?

This device would rest on the ground and the operator would stand on it. With the change in configuration came one huge advantage: they could add larger fuel tanks. The jet belt had burned six pounds of fuel per minute. If the pilot was required to carry the fuel at take-off, he would think twice about taking a longer, heavier flight. And while much of the weight of the jet belt was supported by a stand at take-off, the pilot always had to be concerned about the possibility of making an emergency landing. Coming down hard with a near-full load of fuel in the jet pack would have been a dangerous event for a pilot.

The pilot of a flying platform would not have to lug the unit—or the fuel—on his back. All the device required was a small and efficient engine. Sam Williams knew he had power to spare with the jet belt's WR-19 engine. The only issue was whether such a thing would fly. And if it did, would anyone be willing to buy it?

In 1970, Williams submitted an unsolicited proposal to the United States Marine Corps for a flying platform powered by a Williams engine.[1] The device was named the Williams Aerial Systems Platform, or WASP. The marine corps declined the proposal but was apparently intrigued by the idea of an individual lift device. The marines wondered if a practical device could be made to quickly take a soldier places not accessible by Jeep or helicopter. Ideally, such a craft would maneuver adroitly through a forest. It should be operable by a nonpilot with minimal training and be easy to transport and maintain.[2] The problem appeared to be one of funding: several agencies thought the project was worthwhile but none seemed to have the money to pay for its development. The marines sought Department of Defense funding but were denied. Without the generous funding it had hoped for, the marine corps issued a call for proposals on the STAMP program: Small Tactical Aerial Mobility Platform. The navy and the army joined in; the army version was called the Small Tactical Aerial Reconnaissance System-Visual, or STARS-V.[3]

The STAMP program gave a list of requirements for vehicle proposals. Some of them were well within the parameters of previous work in the field, such as running on a readily available fossil fuel. The vehicle had to be easy to operate and also relatively easy to transport. On the other hand, two new requirements were out of the ordinary: the unit had to be able to fly with two men on board, and the unit had to be able to lift another empty unit like itself and carry it a distance. Presumably this feature would allow one STAMP vehicle to retrieve another STAMP vehicle that had broken down.[4] The proposal gave bidders wide latitude in how they developed a machine to fit the STAMP guidelines, but it offered suggestions. Preference would be given to designs that could go places motor vehicles and helicopters could not go. The machines should be quiet but the proposal used the military definition of "relatively quiet," which meant "no physiological damage" to the operator or support staff.[5]

The marine corps funded several projects under the STAMP umbrella, one of which was the Williams WASP. The WASP was basically a jet engine on a stand with a small step at its base for a pilot. Another step allowed a man to stand on the other side of the engine, facing the pilot. The WASP was a simple, functional design that appeared unfinished due to the visible engine. It served as a prototype for this type of device.

The engine was now called the WR19-9. It weighed sixty-seven pounds, was twenty-four inches long, and twelve inches wide. It was capable of generating up to eleven hundred pounds of thrust, although the designers would not push it that hard in this configuration. The unit with the pilot and the fuel load wouldn't need anywhere near that kind of thrust to lift off.[6] Williams was awarded a contract on September 1, 1972, to build and demonstrate a working model of the WASP.[7] The contract paid the company $1,040,306 for development, with the money furnished by the US Marines. Testing was to be conducted by the Naval Weapons Center.[8]

The WASP was a deceptively simple-looking device. It appeared to be little more than an upright jet engine standing on thin legs with small footholds for the pilot and passenger. The engine drew air from above and blasted it downward through a set of three exhaust ducts. The WASP pilot steered by modulating the thrust between these ducts. The pilot could change directions, turn, and move forward or backward while flying. The WASP was a departure from the other devices Courter had flown, which were controlled kinesthetically. The designers of the WASP had wondered if something this size could be controlled kinesthetically and had decided to use thrust modulation to be on the safe side.[9] This design meant the WASP could not be been flown by someone without some pilot training. Of course, Courter was an extremely experienced and able pilot so it made little difference to him.

In July 1973, Courter flew the WASP tethered.[10] By September Williams announced that the WASP was available to demonstrate taking two men aloft. In December, the military sent observers to Williams and watched the WASP fly, still tethered.[11] It flew fine with Courter on board alone but it was a close call with two men. Williams removed a few items from the WASP, such as an extra fuel tank and a piece of the landing gear, kept the fuel to a minimum, and told Courter to push the engine to the limit. Then it flew with the two men on board. Although the WASP was flying in a slightly modified form, the observers noted that it could fly and deemed the demonstration a success.[12] While it might seem odd that Williams ran the two-man test without knowing that the WASP would not fly immediately, several factors may have come into play. The jet-powered devices such as the jet belt and the WASP needed more thrust to fly in warm air. The temperature may have been higher the day of the demonstration than it was when they tested it previously.

Bob Courter also flew the WASP on a tether on January 8, 1974, for *Popular Science*. An article entitled "Jet flight without wings" featured pictures of Courter flying the WASP by himself and with a passenger. At this point, Courter had flown more than eleven hundred times in the various individual lift devices, and the WASP was just another step in the genre's evolution. *Popular Science* noted that the WASP was specifically designed to carry two men at up to seventy-five miles per hour.[13]

The article noted that the WASP was merely one of the entries in the STAMP competition the marine corps was conducting. A competing company named Garrett proposed taking the fuselage of a Hughes helicopter, removing the engine and rotors, and installing a huge turbo shaft engine to drive a ducted fan system producing 1,050 pounds of thrust. Garrett tether-tested its device and was optimistic it would meet the guidelines of the STAMP program. While discussing whether this vehicle or the WASP would ever be available for the public to purchase, *Popular Science* was typically optimistic. "It's estimated that, in production, such a vehicle would be in the price class of a luxury car."[14]

The STAMP program, and the vehicle proposed by Garrett, showed how the individual lift devices were evolving in different directions. Garrett's two-person vehicle was far too large for any kind of kinesthetic control. Garrett had pitched their design in an unsolicited bid and the Naval Weapons Center had supplied funding for a feasibility study of the device.[15] That report was furnished on March 27, 1974.[16] Garrett proposed a side-by-side configuration for the pilot and passenger. Garrett hoped it would be able to stay aloft for thirty minutes and travel at seventy-five miles per hour. As part of the study, Garrett built the prototype. The company had not free-flown the prototype, but it had conducted successful tethered flights.[17] Like so many other players in this arena, Garrett was overselling the capabilities of the device; what it was building would never fly seventy-five miles per hour nor be able to fly for half an hour.

Garrett took the fuselage of a Hughes light observation helicopter loaned to them by the government and wedged an AiResearch TSE 231 turbo shaft engine into it. The engine was provided to the project by Garrett. The developers pointed out that if the device was manufactured it would utilize a fuselage specially built for the device. The engine was the company's most popular, being used in business aircraft like the Fairchild Merlin III; it produced over nine hundred shaft horsepower and weighed 355 pounds in its usual configuration.[18] In the STAMP feasibility report, Garrett promised it could slim the engine down to 171 pounds; the stripped-down engine would only produce 474 shaft horsepower, however.[19] After all was said and done, the engine would provide a thousand pounds of thrust at sea level on a standard operating day.[20] This should have been an indication of a problem: the report stated that the craft weighed 1,111.8 pounds when loaded with only ten gallons of fuel, a pilot weighing 150 pounds, and a passenger weighing no more than 132 pounds.[21] As described, it would burn a gallon of fuel per minute and probably would not fly.[22]

The Garrett vehicle's engine generated thrust with a ducting system. It pumped air that was split into two side ducts that could point straight down or angle toward the front or back. To take off, the operator would simply point the ducts straight down; to move forward the ducts would aim backward, and so on.[23] Even in a feasibility study, Garrett discovered that its proposed craft had problems. The controlled thrust of the AiResearch STAMP vehicle meant that the pilot had to work controls to orient the vehicle and move it through three dimensions. While the rocket belt and jet belt had been controlled by simple movements of the pilot, the AiResearch STAMP required the pilot to convert his desired direction into the movement of levers and controls. Because the AiResearch STAMP was, in essence, a large hovercraft, it brought a whole new set of control problems to light. If the vehicle was moving forward, simply neutralizing the forward motion would not bring the AiResearch STAMP to a stop. The vehicle would coast unless the pilot actively reversed thrust momentarily to counter the forward motion.[24]

This movement was complicated by the fact that the thrust and control of the vehicle were both to be accomplished with the same ducts. Garrett did testing on its mockup and found that the vehicle was hard to control in this configuration. Garrett proposed more testing and thought that it was probably inevitable that the thrust and control elements be separated and handled independent of each other.[25] This, of course, meant that the STAMP vehicle was going to be even more complex if it ever went into production and would most likely start to resemble a typical aircraft in its piloting requirements.

Garrett built its prototype and tethered it in a hangar. Over a three-day period in 1973, Garrett logged seventy minutes of run time and managed to get the vehicle to hover eight inches off the floor for fourteen minutes with two men aboard.[26] The report did not indicate how much the two men weighed or how much fuel the AiResearch STAMP was carrying. It also included the caveat that the entire flight was conducted within ground effect—essentially, so close to the ground that the vehicle was being partially supported by the cushion of air trapped beneath it as it hovered.[27] Such an effect would disappear if the device was flown just a few feet from the ground. The readers of the report were not sold. The STAMP program was ended on June 11, 1974.[28]

Later, aeronautic experts reflected on the Garrett STAMP prototype and came to the conclusion that it marginally met the requirements of the STAMP program. Its controls were adequate in some respects but not spectacular. It had never flown outside of ground effect but the prototype was probably heavier than it needed to be since it was made from a helicopter fuselage. At 770 pounds, it was heavier than the marines had wanted it to be: they wanted a vehicle that could be transported easily by two men, something in the range of 500 pounds or less.[29]

None of the entrants in the STAMP program generated enough interest for further government funding. Sam Williams, however, refused to give up on the idea. He believed the WASP held promise and he believed the answer lay in making it simpler. Whereas the WASP had controlled its direction with ducted exhaust, its successor would return to kinesthetic control.

12

THE WASP GETS KINESTHETIC CONTROL: THE WASP II

Was kinesthetic control possible on a jet-powered flying platform? The Williams WASP consisted mainly of an upright jet engine on thin legs, with a small platform for the pilot to stand on while essentially hugging the engine as it lifted him off the ground. The designers weren't sure how well kinesthetic control would work on the platform. The rocket and jet belts were both propelled by thrust delivered well above the center of gravity of the pilot. The platform would have its thrust at its base, well below the unit's center of gravity. Would this make the platform prone to tipping over? The Hiller platforms had flown with kinesthetic control but they were also much wider. The WASP had been built with its exhaust gases routed through three nozzles underneath the platform. The nozzles were controlled by the pilot but turned out to be substantially trickier than the two simple controls of the jet belt. Courter thought the platform seemed pretty stable on its own, without his input into the controls. Maybe a platform could also be flown with nothing more than the kinesthetic controls of the jet pack?

The Williams designers decided to build another flying platform and give it just two controls. The right hand control would be the throttle, the left would be the yaw control. The yaw on the jet belt had been controlled by the jetavators placed in the stream of the exhaust; on the WASP II, the yaw would be controlled by vanes placed in the exhaust. The only difference from the jet belt control would be that the exhaust of the WASP II was under the pilot instead of behind his shoulders.

Williams described the WASP II as a "simple, one-man vertical take-off and landing mobility platform, utilizing the thrust of a vertically oriented turbofan engine as its lifting force, designed to transport an individual at low subsonic speeds, and nap-of-the-earth altitudes over water and all types of terrain." Beside the controls, the WASP II was simple for another reason: "the operator is an essential part of the WASP II System which provides the integrating factors for the kinesthetic pitch and roll control of the vehicle."[1] In other words, most of the steering of the WASP II would be done by the pilot shifting his weight and leaning in the direction he wanted to travel.

As for the controls, they were simple too. The yaw control would cause the unit to rotate in the direction the pilot turned the control. If he wanted the WASP to rotate clockwise, he turned the handle clockwise. The throttle was simple, but backward from what many people might expect. To increase the engine speed, the pilot turned the handle counter-clockwise. People familiar with motorcycle throttles might be expected to think the throttle should be turned the other way based upon how the operator's wrist works on a throttle. The handle also had a "positive detent" that would catch the throttle and hold it at an ideal idle speed when it was sitting on the ground, preparing to launch. It would also keep the pilot from inadvertently shutting the engine off while in flight. To move the throttle past the detent in either direction required the pilot to press a button on top of the throttle while turning the handle.[2]

The WASP II was also a relatively simple device to maintain in the field. It would require no more than thirty minutes between flights for the operator to refuel, check the engine oil, and perform the pre-flight checklist.[3] Keeping in line with all of the other individual lift devices that had preceded the WASP, Williams was confident that it could train a nonpilot to fly it. Williams did not overpromise though; the company said it would take two weeks of training.[4] On September 25, 1978, Tank and Automotive Research and Development Command, a group within the military with the unwieldy abbreviated title of TARADCOM, granted Williams a contract worth $1,582,000 to build and demonstrate an operational WASP II.[5]

Williams began testing an engine—dubbed the F-107 when it was used in a cruise missile—to see how it ran when it was mounted vertically. Other than a consideration of how oil settled in the unit when it sat between runs—solved by the installation of a simple check valve—the engine ran just fine.[6]

Williams decided to upgrade some of the internal components of the engine. The F-107s that powered the cruise missiles to targets were not expected to survive more than one flight. An engine failure in a cruise missile was not a good thing, but such a failure would not have a fatal impact on a pilot. The WASP II would have an operator—engine failure was out of the question. Williams engineers went through the F-107 and beefed up everything they could to increase the durability of the engine. Some parts were made stronger, while others were machined more precisely—an unnecessary expense when the engine was used in a cruise missile. The resulting WASP II engine was different enough to inspire a new designation: the WR19-7.[7] Although the original proposal had pegged the engine's output at 545 pounds of thrust, the Williams engineers managed to tweak 640 pounds of thrust out of it by the time they were done.[8] In fact, the engine's output was found to be above the necessary parameter for the military. After crunching some more numbers, they decided to decrease the output of the engine to ensure more reliability. Setting the engine to put out 585 pounds of thrust, it still had turbine blades spinning faster than 61,000 rpm.[9]

The WASP II weighed 235.5 pounds without the pilot or fuel. The vehicle was designed to carry 150 pounds of fuel, almost five times the amount of fuel the jet belt pilot could launch with, but for most flights would carry less. Williams predicted the WASP II would have a flight time in the range of a half hour.[10]

On December 1, 1978, the Williams designers drew what became the final layout of the WASP II design. Knowing that the WR19-7 was the key component, they built an aluminum shell around it that looked like a construction barrel, or perhaps a round pulpit. The tube that formed the body would cover the front and sides; the back would be open to allow the pilot to step in or out of the WASP II. The engine itself occupied the middle of the craft, leaving the pilot straddling it as he worked the

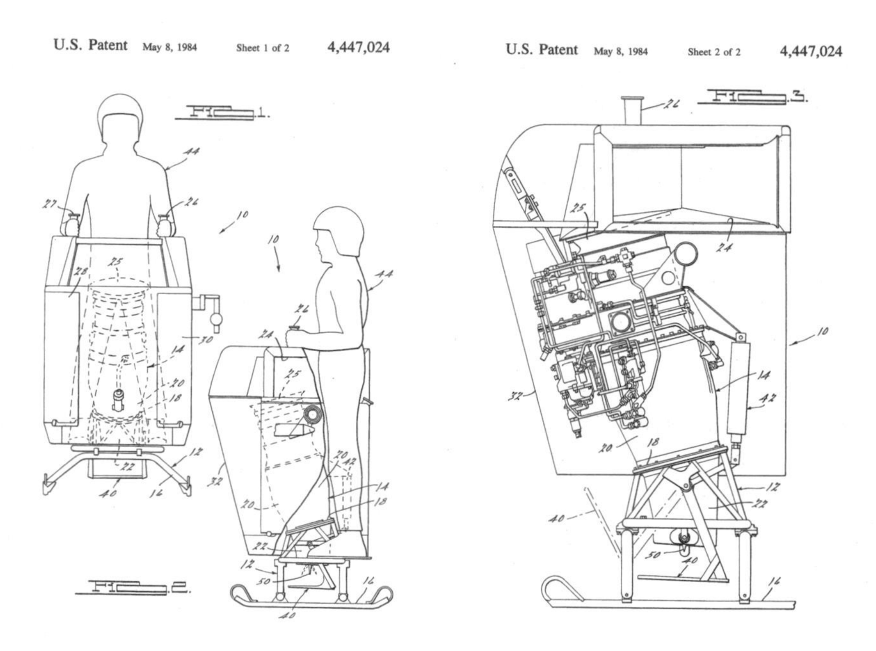

Patent drawing of the WASP II, the kinesthetically controlled flying platform. Control of the aircraft by leaning was described as a major component of the invention.

controls. Perhaps to comfort the pilot, the engine was wrapped in Kevlar so that in the unlikely event of something in the turbine coming loose, it wouldn't escape the engine housing and injure the pilot.[11] The WASP stood on skids like a helicopter, and would give the appearance of the bucket on the end of a cherry-picker when in flight, just without the arm holding it in the air. Others weren't so kind in their descriptions of it. A writer for the *New York Times* described it as "a flying garbage can."[12]

Williams had to fabricate a variety of parts to make the WASP II come together. The WR19-7 engine could not simply have its air intake drawing air from in front of the pilot. A jet engine's intake draws in air with far too much force for a person to simply stand next to it for any length of time. Williams designed ducts that reached over to the sides of the pulpit and drew air in from the sides. The ductwork cost the engine a small loss in performance but it was the only solution.

The WASP II included an emergency recovery system, a ballistically deployed parachute that could be launched by the pilot in case of an emergency. The parachute and charge were located in the front of the body of the craft and to launch it the operator had only to press a panic button on the top of his hand controls. Because the chute was deployed by an explosive charge, it was vital that the chute not be deployed accidentally. Williams included a warning in the system specification that the arming switch for the parachute deployment system be in the disarmed position and only armed in the moment immediately before the pilot lifted off in the vehicle.[13] Upon boarding the WASP II and preparing to take flight, the pilot would clip the lines from the parachute to a harness he was wearing so that the parachute would lower him to the ground in case of an emergency.[14] In such an instance, the WASP II would simply be left for gravity to deal with.

On September 28, 1979, Williams had a fully assembled WASP II ready for gimbal testing, without a pilot.[15] Williams engineers mounted it on a device that held it in place but allowed it a small range of movement to test its stability. After running the WASP II on the gimbal with ballast, they then tested the ballistic parachute system.

On October 25 and 26, 1979, engineers hoisted the WASP II off the ground with a one-hundred-foot crane. They dropped the WASP II and fired the parachute. After four successful test drops, the recovery system was deemed effective.[16] There was one small problem though. Even a ballistic parachute takes a few moments to inflate. From one hundred feet, the system worked fine. Below seventy-five feet or so, the system would not be effective. A pilot would always have to be concerned about problems that happened more than a couple of feet off the ground but below seventy-five feet.[17] After the successful test, they were convinced the vehicle was safe enough to test with a live pilot. Bob Courter got the call.

On February 12, 1980, Courter tested the WASP II, serial number 101, on a gimbal and confirmed the vehicle's stability.[18] On February 20, Courter made his first tethered flight in the WASP II. As he flew it on the leash, he made mental notes and talked with the engineers. They decided the throttle controls needed tweaking because they weren't sensitive enough. They also decided that the jet blast downward—it had been angled backward slightly at 17.5 degrees—needed to be moved slightly forward so the angle became 15.5 degrees. Each improvement made the WASP a bit more stable and Courter soon found himself mastering another space age flying device. Courter flew the WASP II tethered forty-five times between February 20 and March 29, 1980.[19]

On April 17, 1980, Courter flew the WASP II without a tether for the first time.[20] He flew it for eighty-eight seconds and put it back down. It was a short flight by some standards but still four times longer than the rocket belt ever flew. More flights followed and soon Courter had a flight where he spent twenty-four minutes in the air.[21] Flight lengths often ranged in the area of twenty minutes, depending on how much fuel they chose to put in the machine and how warm it was outside.[22] The kinesthetic control worked. Courter could get the WASP II to change directions, move, slow down, or stop, all by shifting his weight. Williams built a second WASP II as a backup; eventually, the company built a third as well.[23]

Courter started racking up airtime in the WASP II and finding out about longer flight times in an individual lift device. He noticed a problem when flying too high. He had to gauge his speed by looking at the ground and landmarks he was passing over. The higher he got, the harder that task became. "Right now we have only flown it at heights of 60 to 70 feet, because at over 100 feet you lose visual cues, and it's hard to tell how fast you're moving up and down,"[24] he told a writer. The WASP II did not have an airspeed indicator, altitude gauge, or any other onboard instruments—it didn't even have a fuel gauge—mainly to keep the weight down. *Popular Science* assured its readers that later models would carry the necessary gauges.[25] They asked rhetorically if the WASP II might be "The ultimate in personal flying machines?"

The army contract required Williams to demonstrate the WASP II to the military, at which point a decision would be made about the device's future. Two demonstrations were made. At one, General Carl Vuono watched as Courter hovered effortlessly around the sky, disappearing behind trees and reappearing unexpectedly somewhere else. The WASP II was loud enough that observers always knew it was nearby, but the vehicle's ability to pop up and disappear was unlike anything anyone had ever seen before. The general reportedly turned and said to the man next to him, "We need something like this."[26] He was a mere general at the time; a few years later he would be the chief of staff of the United States Army. Williams finished the project under budget and two months ahead of schedule.[27] Officially, the contract was declared completed on July 8, 1980.[28]

Williams International—it had changed its name from Williams Research in July 1981—submitted a patent application for the WASP II on February 8, 1982, entitled simply "Airborne Vehicle."[29] The patent listed only one inventor: Sam Williams. The patent was only five pages long, three of which were drawings of the device. The primary claim made on the application addressed the manner in which the vehicle was steered while in flight. "While airborne vehicles that rely only on thrust for ascent are known, movement in a horizontal plane has not heretofore been achieved by a change in the center of gravity of the vehicle relative to vertical thrust of the engine." Later, "In this manner, highly effective kinesthetic control of the vehicle is achieved."[30]

In 1981, Williams had looked for another pilot to back up Courter, who was the only person trained to fly the WASP II. A pilot named Mark Voss, who also had a degree in engineering, applied for the job. After some basic back and forth with the Williams personnel department, he found himself alone with Dr. Williams, the man who founded Williams Research in 1954 and now ran a world-renowned aerospace company. Voss later said, "He was a man of very few words and you could sense the intelligence. Whenever he spoke or asked a question, it was extremely compact. When the words came out, you knew they would count. You paid attention when he talked."[31]

Voss found out that one of the reasons he was being recruited was that Courter had been putting on weight. Before each WASP II flight, the engineers would check the weather because air temperature played such a pivotal role in the engine's performance. Then, they would weigh Courter. Using those two factors, they could calculate the vehicle's expected flight performance, including how long it could stay aloft. While the engineers were struggling to get longer flight times out of the WASP II, Courter's extra weight was countering their efforts. At his heaviest, Courter sometimes would not be able to hit fifteen minutes of flight time. Voss stood five foot four inches and only weighed 117 pounds; the engineers knew that a simple change to a lighter pilot would go a long way toward extending WASP II flight times. He was hired in as a project engineer specifically to support the WASP program, but also to train as a backup pilot when the time came. Courter's feelings weren't hurt. Voss later said, "We were getting so penalized by his weight that a conscious decision was made by Bob [Courter] in order to help keep the program moving along to remove the ballistic parachute and to start flying without that on board."[32]

The engineers who worked on the WASP II also discovered a bizarre acoustic anomaly. At certain heights, a whooping sound like a car alarm was created due to the harmonics of the exhaust and the speed it was traveling out of the bottom of the craft. The sound appeared between two and five feet of altitude. Voss described, "It made an incredibly loud whooping noise. We tended to get out of that really quickly." Even though the noise could be heard several hundred yards away, the Williams engineers didn't bother addressing it. Other than calling attention to the WASP II, the sound didn't actually affect anything.[33] "It was about a two or three hertz tone, very high-pitched. It had absolutely nothing to do with engine tone. It was just an acoustic standing wave that was created between the engine duct and the ground that sounded like a car alarm," Voss said.[34]

The next step was a concept evaluation program where Williams would prove the viability of the WASP II as something an average soldier could learn to use. The military provided another million dollars, this time to get the unit confirmed as airworthy and then, assuming that was accomplished, to train three nonpilots to fly the device.[35] Bell had trained nonpilots to fly the rocket belt but had picked their own test subjects. This would be the first time that someone other than the craft's manufacturer would select pilot-trainees for an individual lift device.

The preliminary airworthiness evaluation was conducted by Major Donald L. Underwood, United States Army, a highly decorated helicopter pilot and graduate of the navy test pilot school. Underwood came to Williams headquarters in Walled Lake, Michigan, to be trained to fly the WASP II by Courter. Over the period from October 4, 1981, to March 29, 1982, Underwood took the controls of the WASP II, first on the gimbal and then tethered.[36] Williams employees watched Underwood struggle with the controls and became convinced he was not the right man for the job. Voss said, "The thing, frankly, scared him. It took a long time to make him comfortable." Underwood's slow learning curve cost Voss his job as backup pilot. "It took so long to make him comfortable; the contract had an allowance for so many number of engine overhauls. Underwood had ended up preventing me from being trained because the priority was to get his sign-off on the WASP II for training for the three sergeants. As a result of the excessive amount of time it took to get him trained, we became concerned that we wouldn't have enough budget to cover the engine time that I would use in training. That was a disappointing day for me, the day that decision was made."[37] Finally, after fifty-one practice runs, Underwood flew the WASP II untethered eight times. Courter was convinced that the WASP II scared Underwood. He had shown Underwood how he could fly it at speeds over sixty miles per hour and easily to heights of a hundred feet or more. Underwood puttered around in the WASP II, never going faster than fifteen miles per hour and never over fifteen feet from the ground. According to Courter, "He was scared to death to fly the thing."[38]

Underwood flew the WASP II in calm weather and in a little bit of wind as well.[39] For the WASP II to be flown by others in the military, it needed an airworthiness certificate. Underwood reported his experience and findings to his superiors. Then, he issued the requisite airworthiness certificate. He had his reservations, though.[40]

During Underwood's test flights, he found a few things he didn't like about how the WASP II flew. Williams had placed an exhaust deflector under the vehicle so that the exhaust gases wouldn't blast directly into the ground beneath the jet engine while it sat idling before take-off. Underwood found the deflector too hard to flip into and out of position and asked project engineers to just remove it, which they did. He also didn't like that the buttons for the emergency recovery system were on top of the controls, especially one button that sat next to another that allowed the throttle to travel past the detent differentiating between ground idle and full throttle. He asked that the system simply be disabled during his tests.[41]

Underwood issued his findings in a Preliminary Airworthiness Evaluation in June 1982.[42] He identified nine "deficiencies" and eleven "shortcomings" that would need to be addressed before the WASP II went into full production for the army. That is, if it ever went into full production.

He felt a few of the issues needed to be addressed before the nonpilots were trained to fly it. Deficiencies needed to resolved first; shortcomings could be taken care of later.[43] Because of the shortcomings Underwood found with the WASP II, it was issued a "limited airworthiness certificate."[44] Army personnel would be able to fly it but only under limited conditions.

Although Underwood's list seemed lengthy, most of the problems were easily resolvable or were simply flight characteristics that had been anticipated and were due to the nature of the craft. He complained that the WASP II had limited performance when the temperature was over sixty-eight degrees Fahrenheit. He thought the unit was harder to balance than it should be, notwithstanding Courter's ability to zip around effortlessly in it. He thought it was too loud at take-off, hovering, and landing. Underwood complained that when it was ten degrees or colder, his hands got too cold. He thought the unit was too sensitive to the position of the pilot—even though the device was designed to be maneuvered kinesthetically. A pilot couldn't get out of the vehicle one-handed; presumably, it took two hands for the pilot to undo his connections to the emergency recovery system. There were the two issues with the exhaust deflector and the emergency recovery system triggers that he had disabled before his own testing.[45]

Everyone knew the WASP II would require more throttle when operating in warm air. In the System Specification for the WASP II, Williams had included a graph entitled, "Temperature Effect on System Endurance," which displayed engine performance at temperatures between thirty and ninety degrees Fahrenheit. The line on the graph was perfectly straight and predictable as performance declined with increasing temperature.[46] The upshot of this phenomenon was just that the device got better fuel economy in cold weather and lesser fuel economy as the temperature climbed. Either way, the designers knew about it before the WASP II was built and had told the army this was the case. It was also an issue that had been observed with the jet belt.

Some of Underwood's other criticisms seemed nitpicky and later would draw suggestions that he had been intentionally overplaying deficiencies with the vehicle. He said that the WASP II "could be safely flown throughout the limited flight envelope with minimal but adequate control margins." He then complained that "extensive pilot compensation" was needed.[47] The vehicle was designed to be kinesthetically controlled; how was that different from "extensive pilot compensation"? It is worth noting that Underwood had not flown any of the other individual lift devices which were kinesthetically controlled in flight. As a helicopter pilot, he understood flight characteristics of machines with elaborate controls. But helicopters weren't sensitive to small things, like the pilot merely shifting his weight or moving his head. Perhaps it was this unfamiliarity that fueled Underwood's criticism of the WASP II.

The army proceeded with the experiment to see if a nonpilot soldier—the military called them "non-rated Army operators"—could be taught quickly how to fly the WASP II.[48] At Fort Benning, the United States Army infantry board showed a picture of the WASP II to some soldiers and asked for volunteers. A staff sergeant named Ray LeGrande Sr. saw the picture and was intrigued. LeGrande had joined the army in 1971 at the age of twenty and had come from North Carolina.[49] Along with the others who volunteered, LeGrande was interviewed and told a little bit about the WASP II. He, Raymond Fitzgerald, and William Duval were chosen for the program. They were told they would be taught to fly the WASP II and if the program was adopted by the military, they might become the flight instructors.

That part was what appealed to LeGrande the most. He had been an airborne instructor and taught people to jump out of airplanes. He had well over five hundred jumps including some at night. He had jumped five or six times a week on occasion with the 82nd Airborne Division at Fort Bragg and never had any trouble teaching people how to get over their fear of heights. The volunteers were told they were guinea pigs to "see if regular Joes could fly it." LeGrande did not know the other soldiers before the program started. The other two were not from the Airborne Division, they were just regular GIs he met in the WASP II program.

After they were selected, they traveled to Walled Lake, Michigan, for what was supposed to be a three-week trip. They got an apartment near Williams and a government car to drive over for training. First they had several weeks of classroom training in a trailer. The men sat through lectures by Courter on all aspects of the WASP II and individual lift devices and were given written tests. It was winter, and it seemed very cold to the soldiers who had just come up from Georgia. The weeks added up. Eventually, ten weeks passed before they were ready to step into the WASP II for the next chapter of training.[50]

After passing the written tests, they moved to flight training. LeGrande was amazed at how simple the operation of the WASP II was. Later, he compared it to the modern device steered by the balance of the rider: "If you could ride the Segway, you could fly the WASP."

First, the trainees "flew" the WASP II on the gimbal apparatus. The device was powered but held in a rig that allowed it to move about in place, forcing the pilot to maintain control as if in free flight. It was difficult at first. Courter was watching and overseeing. If the operator lost control, Courter could shut it down. Although the experience was a new one for these inexperienced operators, they each managed to settle the craft down with just a day of gimbaled flight. After learning to balance the WASP II while gimbaled, Courter moved them over to the tethers. There they had more freedom but still a proverbial safety net. Each man logged more than fifty flights, a total of about three hours, with Courter standing by to offer guidance, yelling over the roar of the exhaust.[51]

Once Courter was convinced the three could fly the WASP II, he moved the show to Fort Benning. The contract called for Williams to furnish two WASP IIs; both were brought there. One they flew and the other was a backup, although they never had to use it. They commandeered a remote hangar surrounded by a large paved area the size of several football fields. The portion of the base was far from prying eyes and usually reserved for helicopter training. Courter ran the three through some more gimbal practice and then let them fly the device on a tether set up on a crane. LeGrande had been doing the best in training, so Courter let him fly it first. "It was lovely," he remembered thirty years later.

LeGrande noticed something about the WASP that the engineers had known was going to happen and was related to Underwood's complaints. The WASP II ran very well in cold weather and seemed to have better lifting ability in the cooler air. In the warmer climate of Georgia, LeGrande noticed it needed more throttle. The warmer air was not as dense, which resulted in the need for more throttle to get the same amount of air moving through the engine as they had gotten in the colder climate of Michigan. LeGrande and the others spent a few weeks flying the WASP II tethered to the crane.

Courter announced it was time to fly the WASP II without the tether. LeGrande went first and soon found himself taken with how easily the machine flew and how fun it was to fly. Without a tether, LeGrande had no problem maneuvering; he flew around the large space at the edge of Fort Benning and enjoyed the view from his flying platform. All of the flights the men took were videotaped. On a tape LaGrande has of all his flights, Courter can be heard saying, "Ray, you're too high!" He was at fifty feet. Without an altimeter, it was easy for a pilot to drift higher without realizing it. Of course, it was an easy matter to bring the WASP II back down. It could do sixty miles per hour but LeGrande never took it much faster than thirty. No one ever crashed the WASP, but one day Duval was flying it and lost control of it for just a moment; it swung like a pendulum as he overcorrected. He got it back under control and landed it. He took the rest of the day off. That was the closest they ever came to crashing one.

The WASP II was also loud. When asked to describe how it sounded, LeGrande could only say it sounded like a jet engine. There was no other way to describe it, because that's what it was. When it was in the air it seemed less loud, particularly to the operator who was above the engine. During this testing there was not a lot of debris kicked up because the area they were using was so clean. The prop wash wasn't bad—"You'd get some," LeGrande said, but it was no worse than from a helicopter.

As soon as they mastered the control of the WASP II, Courter reminded them they were going to have to fly within the limits of the Preliminary Airworthiness Evaluation: no more than fifteen feet off the ground and no faster than fifteen miles per hour. More than once, Courter had to tell the men to bring the WASP II lower or slow down because they forgot to follow the rules. LeGrande remembered hearing Courter yell at him, "Slow down! Slow down!" LeGrande thought the fifteen/fifteen limits were unnecessary, but he also knew how the army was with regulations that seemingly made no sense to the soldiers.

Simply learning to fly the WASP II safely was not enough. Courter explained to the men that they would have to navigate the craft through a course. It wasn't an obstacle course, it was simply a design on the ground. The WASP II pilots would have to take off, go left, go right, go forward, do a 360, go backward, and then land, following an out-and-back set of lines drawn on the ground. They had to negotiate the pattern successfully to show the viability of the WASP II.

LeGrande remembers that the course was easy. They could have done it better if they were allowed to fly a little bit higher than the fifteen-foot ceiling they were held to. The lower height meant there were slight issues with the jet blast down from the WASP ricocheting back up and causing a bit of turbulence when the operator got too low. LeGrande felt that the tests were sabotaged. That is, the test they were given was something they could master quite quickly, but the army wasn't asking them to show what the WASP II could really do. "They were trying to hold us back."

Thirty years later, LeGrande remembered flying the WASP II with fondness. It "was unbelievable. It's a wonderful feeling. Once you get control of it you can do almost anything with it. There were still some things we hadn't even done with it. It was a magnificent thing. You take off vertically, you can just hover, you can do a 360, go left, right, backward. Go straight up and just stop. Just stop and look around— it was just wonderful."

While LeGrande and the others became quite adept at flying the WASP II, none of them compared to Courter, who had by now logged flights with rocket belts, jet belts, the WASP, and the WASP II. Courter would fly the device anytime someone official wanted to see it or when Williams wanted to show what the craft could really do. Courter made it look easy; he flew it effortlessly and most people who saw him fly it left with the impression that it almost flew itself.

The army set aside two weeks for the WASP II demonstrations and allowed the press to come in and get a preview before the testing started. Newspaper reporters described the "army testing a one-man, jet-powered flying platform that resembles a fictional aircraft in the Dick Tracy comic strip." Courter fired up the WASP II for the reporters and did not disappoint. "With a deafening roar, the WASP climbed to about six stories, ducked behind a row of trees, performed various maneuvers, and came to a sudden stop 100 feet in the air."[52] Reporters walked around the WASP and photographed it. LeGrande and the others posed in the machine and talked with reporters after. Glowing stories appeared in the papers about the army's new secret weapon. According to the paper, the tests were going to cost $50,000 and $3 million had already been spent developing the WASP II.

After LeGrande and the others had all demonstrated how they could run through the course laid out for them, Sam Williams and the rest of the engineers from Williams answered questions for the army officials. LeGrande and the others returned to their pre-WASP II duties. He thought he would hear something soon, but he never did. No one ever called him to tell him whether a decision had even been made. He assumed that since the demonstrations had all gone off without a hitch and the press had been called in to document the triumphant experiment, it was just a matter of time. He went back to his unit and waited. Today, he is "still waiting."[53]

In May 1983 Williams was invited by the army to bring the WASP II and participate in some battle simulations in Yakima, Washington, at Fort Lewis. The army was evaluating new infantry equipment. Courter took to the air using the WASP II as a reconnaissance vehicle and observed and reported on the activities of an enemy force under simulated combat conditions. Unlike before, the military did not impose conditions on Courter, the former combat pilot. He performed three combat missions and on one occasion reached forty-seven miles per hour while zipping over the battlefield. To allow for better maneuvering and more speed, he ran the WASP II without a full fuel tank. As a result the flights were no longer than five or six minutes. Afterward, the army said the WASP II had performed well but the short flights would not have been adequate on a real-world battlefield. The noise generated by the device hadn't been an issue either but they wondered about the heat signature of the WASP II.[54]

This test resulted in army officials concluding that the WASP II was impractical as a battlefield tool. They believed its flight time was too limited and the lack of armor and weaponry on the vehicle made it vulnerable. Of course, adding those things would have limited the flight time even further. They also said that they doubted the ability of a WASP II pilot to fly the machine safely, avoid ground fire, and perform reconnaissance tasks efficiently and safely.[55]

What the army did not consider was that while Courter was an exceptional pilot, he had gained quite a bit of weight since his days as a rocket belt pilot. Even during the WASP II program, his size had increased. Toward the end of the program, the engineers had a difficult time tweaking ten-minute flights out of the craft with Courter at the helm. "That's when people really started to get frustrated, when we got under ten minutes," one WASP II engineer said later.[56] Perhaps the program would have impressed the army more if the WASP II had been flown by a lighter pilot who could have achieved a longer flight time.

Williams believed the vehicle's perceived shortcomings could be resolved. This unit was, after all, just a prototype and Williams had been studying the vehicle as closely as the military. With some redesign and improvement, the WASP II would overcome the problems identified by the army. Williams submitted a proposal to deliver twenty updated WASP IIs. The entire program, including redesign, tooling, manufacturing, assembly, and delivery would cost $35 to $40 million. From that point forward, the cost per WASP II would come out to only $250,000. The army declined, citing the high cost and the short flight time.[57]

A final problem was also the skill level needed to fly it. Courter had made it look easy, but he was an experienced combat pilot and licensed flight instructor. LeGrande and his cohorts, although not pilots, were probably more skilled than the average GI who walked into America's recruiting stations at that time. The army was hoping for a device that required no more talent or training to drive than a motorcycle or a Jeep.[58] Later, experts who studied the program pointed out some other issues that may have occurred if the WASP II had gone into production. While it could carry quite a bit of fuel, it also burned it quickly. The weight of the craft dropped as the fuel was burned. As it lost weight, it became lighter, faster, and easier to maneuver. Would pilots have trouble learning to compensate for the decreasing weight they experienced each time they flew?[59] The exhaust from the WASP II's engine was 620 degrees Fahrenheit and blasted at fourteen hundred feet per second. Could the exhaust catch dried vegetation on fire? What about debris blasted off the ground? Could it get sucked into the air intakes on top of the WASP II?[60] This situation had occurred a couple times during testing but had never caused the vehicle to fail. Still, all these questions kept popping up during the program.

By 1983, government agencies had spent $7,760,000 developing and testing jet-powered individual lift devices such as the WASPs and the STAMP prototype by Garrett.[61] But Williams refused to throw in the towel just yet. The army retained possession of the two WASP IIs that were procured in the contract; Williams had enough parts to build a third. The company assembled another device and began pitching the idea of the WASP II for civilian use. In some respects, it was not unlike Bell pitching the rocket and jet belts to the public, but Williams would not do widespread demonstrations of the WASP II. Instead, its salesman hit the road and began trying to drum up business from the most likely suspects. Would law enforcement be interested? How about fire and rescue? The third WASP II was painted in civilian colors—the other two had been painted in olive drab—and had the name X-Jet painted boldly across its front.[62]

It wasn't just a matter of finding people willing to buy one or two WASP IIs. The WASP II as it was delivered to the US Army had used an engine originally designed to power a cruise missile for what amounted to a single, one-way flight. The requirements of the cruise missile program had sought to bring the cost of the engine down as much as possible and endurance was not an issue. Further, the cruise missile application did not require the engines to be rated as highly as engines that powered vehicles that carried people. The WASP II engines had been rebuilt to be able to fly perhaps twenty hours between teardowns. A civilian version of the device would have needed an engine with figures approaching ten times that—more on the magnitude of two thousand hours between overhauls. "Two thousand hours is a pretty typical expectation for the life of an aviation engine," one engineer later said.[63] Williams could design such an engine, but it wasn't interested in doing the work unless they found a customer willing to buy the machines. To develop an engine for a civilian X-Jet might have cost as much as $2 to $2.5 million. Although Williams found some interest in the machines, no one wanted to commit the money to allow them to take the X-Jet to the next level.[64] The project was quietly shelved.

In 2000, a writer for *New York Times Magazine* did a story on jet pack technology and asked Williams International about the WASP II and why it was not available today. "Today the company is curiously reticent on the subject though jetpack fanatics continue to call with inquiries on a regular basis; a spokesman for Williams dismisses the issue as something from 'the very old past.'"[65] A Williams spokesman told the writer that the WASP II—the X-Jet—was probably most hobbled by its short flight time, not compared to other individual lift devices, which it could fly circles around quite literally, but to other flying devices like helicopters. That writer concluded that the failure of the WASP program suggested that jet packs would likely never be available. "Though it saddens me to think so, maybe the answer is that jetpacks no longer exist in the future but behind us, receding into the past: a device for memory instead of flight."[66]

13

THE ROCKET BELT RETURNS

Contrary to the writings of the *New York Times*, jet packs were never forgotten or abandoned. They just slipped off the public's radar because corporations such as Bell and Williams were no longer actively developing them. However, an active interest in the devices had grown over time. Individuals, using their own money, built and operated rocket belts. Some spent years and tens of thousands of dollars in their quest to fly.

The same year that Bell grounded their rocket belt flying team, a Hollywood moviemaker named Nelson Tyler built his own rocket belt. He saw Suitor fly the Bell Rocket Belt at Disneyland in 1965 and decided he wanted one. He had built amazing devices for filmmaking and thought he could probably build a rocket belt if set his mind to it. He asked Suitor if he could get his picture taken with Suitor while he was still wearing the belt. Tyler held a measuring stick in his hand so that he could extrapolate measurements for the belt from the photographs. He then set out to copy the Bell design as best he could based upon his photographs and other photos and drawings he could get his hands on. He ran into a snag when he got to the throttle valve; photographs of the belt didn't reveal the details and nuance of the device. Doing further research, he found the name of the National Water Lift Company, who had manufactured and sold the original rocket belt valve to Bell. People at National Water Lift knew the dimensions and measurements of the parts and were happy to help Tyler out.

Tyler sold his car to raise some of the funds. It cost him around $15,000 to build what he called the NT-1.[1] He fired the engines with the device tethered, but he couldn't seem to get the belt to work. Tyler discovered one of the facts of rocket science that others would learn after him: the belts can be finicky and it takes specialized knowledge to get them to fly. Tyler knew he had to consult an expert.

He contacted Bill Suitor for help in fine-tuning the belt and Suitor was happy to oblige. In fact, Suitor had only recently left the employ of Bell and thought he might never see a rocket belt fly again. Suitor hopped on a plane to California and was soon firing Tyler's rocket belt's engines. He confirmed that it appeared to be operational. Even though the belt had not been manufactured by Bell, Suitor believed it would safely fly. He put it on and took it for a flight. It was just like flying the Bell belt. Tyler hadn't built the belt for any particular purpose: he was simply fascinated by the belt and had decided to act on the impulse.[2] Tyler would be the first of many individuals who, after seeing the belt demonstrated someplace, would be compelled to build one of their own.

Tyler—after training from Suitor—was soon flying his own rocket belt. Through his Hollywood connections, he managed to make some commercial appearances with it. He starred in a Canadian Club print ad wearing the belt and he booked some movie appearances with Suitor at the controls. The *A-Team*, *The Six Million Dollar Man*, and *Newhart* introduced the rocket belt to a new generation of viewers.[3] Others would take note of the money-making possibilities of the rocket belt. Some would find great success; some would come to a tragic end.

Clyde Baldschun, who had booked the Bell Rocket Belt into all of the state and county fairs, told Tyler he'd be happy to find places willing to pay for rocket belt appearances. Just because Bell was no longer in the game didn't mean the audiences wouldn't be there. In the next fifteen years, the Tyler rocket belt flew 120 times, most with Bill Suitor at the controls. Tyler did fly the device a couple of times but was content to let others do most of the flying for him.[4]

In 1982, Suitor flew the Tyler belt at the Knoxville world's fair but got angry when he found out that Baldschun had been cutting corners on insurance, something Bell had always been careful about maintaining.

Suitor decided to retire from rocket belt flying again. Around this time, Tyler hired another pilot, Kinnie Gibson, a stuntman who had worked with Chuck Norris. Suitor helped train him to fly the rocket belt. Gibson flew the belt on a tether thirty-one times and then flew untethered.[5] Soon, he was flying the Tyler rocket belt before huge audiences as well.[6]

Shortly before the 1984 Los Angeles Olympics kicked off, Tommy Walker, the man in charge of special effects for the opening ceremonies, contacted Baldschun about hiring the Tyler rocket belt for a starring role in the extravaganza. Baldschun and Tyler were excited about performing in front of such a large audience but were concerned when Walker started explaining how precise he wanted the flight to be. It had to be choreographed to fit snugly with numerous other elements of the event at the Los Angeles Coliseum. The pilot would have to launch on cue from one specific spot, fly over to another, make a turn, and then land on a target, all with split-second accuracy. Baldschun and Tyler wanted Suitor to make the flight, but they didn't tell Gibson that.

While Gibson was calling his friends and family to tell them he would be in the opening ceremony of the Olympics, Baldschun and Tyler were trying to convince Suitor to come out of retirement for one last big flight. Suitor was hesitant. He was in the middle of a big job at home and he couldn't get time off from his real job. But the promoters were relentless, and Suitor finally gave in.[7]

On July 28, 1984, Suitor put the rocket belt on for one last grand finale. Standing on the top step of the Coliseum, Nelson Tyler helped him with the rocket belt and double-checked everything for the high-profile flight.[8] He rocketed over a live audience of ninety-two thousand people. Billions of viewers around the world—*billions*—watched the flight on television. Suitor spent only seventeen seconds in the air, but the flight was the highlight of the opening ceremony. The promoter was paid $7,500 to book the flight; Suitor was paid $1,000 and travel expenses. Later, he found out that Baldschun had not bought an insurance policy for the flight, even though the two had discussed it at length, particularly after the argument they had at Knoxville.[9]

Suitor's Olympic flight put the rocket belt back in the headlines. Many people were shocked to discover that the belt was owned and had been built by a private individual. Tyler got offers from people who wanted to buy his belt, and he eventually sold it to an amusement park in Sweden for $250,000. The park hired Kinnie Gibson to make a series of flights there.[10] Afterward, Gibson bought the belt and returned to America to start a business flying it.[11] Gibson soon found out that maintaining it and finding fuel created a new set of problems. Bottled nitrogen was expensive, so Gibson experimented with using compressed air as a propellant, as scuba divers do—but the impurities in the air degraded the performance of the belt. Instead of twenty-one seconds of flight time, the belt sometimes sputtered out before that. Once, while shooting a commercial, he was forced to make an emergency landing and scraped his knees.

Hydrogen peroxide was also becoming harder to obtain. The only supplier he could find was in Germany, but the stuff they sold contained a stabilizing agent that, over time, corroded the catalyst section of his rocket belt's engine. A flight in Philadelphia ended in a crash as a result. He broke his knee and wound up hospitalized. He sued the company that sold him the hydrogen peroxide for not telling him about the additives in the fuel. The case was settled, but the company decided to stop selling hydrogen peroxide to the public. Gibson began making his own fuel.[12]

Gibson hired an old friend of his named Brad Barker to help him with his rocket belt company. The two had met while working together at an insurance company in Texas in the 1970s. Barker had been enamored with the belt ever since he first saw one fly in *Thunderball*. He helped Gibson repair the belt after the crash in Philadelphia and often accompanied Gibson on his trips to fly the belt at events. In essence, Barker was Gibson's ground crew.

Eventually Barker and Gibson had a falling out and Barker returned to Texas. Barker had seen the rocket belt fly many times and knew how much money Gibson was making. Before he left, he had seen Gibson win some lucrative contracts. Gibson had cut one spectacular deal with Michael Jackson, who had hired Gibson to fly off stage at the end of Jackson's shows for one tour. The flights earned Gibson $25,000 each and added up to almost a million dollars for the tour.[13] He also signed a contract to provide a series of twenty flights at various events at Disney World.[14] It looked to Barker like Gibson's rocket belt was a money-making machine.

In 1993, Gibson ran into a problem with his belt. While flying it tethered, the throttle malfunctioned. Once the fuel ran out, the throttle would not return to its starting position: the valve was stuck. Gibson began looking around to find out who might be able to help him with repairing or replacing the valve, and he wound up in contact with the people from National Water Lift. It had been decades since they had built the original valve for Wendell Moore at Bell and the men who had designed it no longer worked there. No one knew anything about it. The company tracked down an old-timer named Bob Kutsche who had retired, hoping he might remember something about the valve. Kutsche hadn't worked on the valve himself, but he knew who had. Many of the key players who had designed the valve had passed away, but Kutsche volunteered to come back to National Water Lift and help Gibson get his belt flying again. Soon they found the drawings—dated 1961—for the valve and located the vendors who had made the hard anodized aluminum parts. Gibson brought the frozen valve to Kalamazoo for the engineers to look at.

Kutsche went to work. He found one engineer who had worked on the Bell valve and soon they had replicated the original valve from the Bell belt. Kutsche suggested to Gibson that he might want to have them manufacture a couple of extra valves so that he would have spares in case he needed them in the future. Recognizing the opportunity, Gibson agreed and National Water Lift made and tested four valves for him. "So, he's set for life," Kutsche said later. Gibson was so grateful to National Water Lift that after he reassembled his rocket belt with the new valve, he came back to Kalamazoo and gave a rocket belt demonstration at the plant.

While Gibson resolved his valve issues, his former assistant Brad Barker was planning to enter the rocket belt field. Barker would eventually have a flying rocket belt—only the second privately owned one to fly—but the story of that belt would involve kidnapping, murder, and exactly one public flight.

14

THE PRETTY BIRD SAGA

Kinnie Gibson's former assistant Brad Barker had formed the American Rocketbelt Corporation in 1992 with a partner named Larry Stanley.[1] Stanley was a man who seemed to have access to money and had dealt with both Barker and Gibson previously. Interestingly, Barker would later claim that Stanley had stolen his plane before they had embarked on the rocket belt project, and had also stolen Gibson's rocket belt while Gibson was filming a movie in the Philippines. Even though Barker had never resolved the plane issue, and claimed to have been in an ugly physical confrontation with Stanley over the stolen Gibson rocket belt, he felt he could make a go of the business with Stanley as half owner.

Barker had photographs of the Tyler rocket belt but little else to go on. He started doing what he could to gather the necessary parts for the device. He found an engineer named Doug Malewicki who said he would help if Barker and Stanley signed a release stating that they would not hold Malewicki liable for anything that happened to them while working with the belt. The two signed.[2] Malewicki understood how dangerous rocket belts could be; he just had no idea that the rocket belt was the least dangerous element of the American Rocketbelt Corporation.

A friend of Barker's named Joe Wright owned a car audio store in the area, and he offered to let the young company use his garage facilities to work on the belt. He would defer rent charges until they started making money. They began assembling a rocket belt that they planned to call the RB-2000.[3] Barker and Stanley decided to formalize their agreement and met with Stanley's brother, who was a lawyer. They drew up formal corporate documents, agreed to issue stock split evenly between the two of them, and gave themselves titles. Barker was president of American Rocketbelt Corporation; Stanley became vice-president, secretary, and treasurer. The corporation's only asset was the in-progress rocket belt and they figured by now they had invested almost $200,000 in the venture, almost all of which they had borrowed from their mothers.[4] Joe Wright, their landlord, was named marketing director. He may have also been promised a 5 percent cut of the profits.[5]

In the fall of 1994, their rocket belt was complete. The belt was similar to Gibson's, but Barker had decided to try to do something about the short flight time. He fitted their belt with larger fuel tanks. The unit was heavier, but Barker assumed the extra fuel would make it fly longer. With a video camera recording, they fired its engine while it was mounted on a stand in Wright's shop and it appeared to work just fine. The problem was that none of them had ever flown a rocket belt before, and Barker couldn't ask Gibson for help. Barker had seen one flown many times but that was not enough for him to strap one on. Barker knew that Bill Suitor had helped train Gibson; he decided to track him down and see if he was willing to test-fly their belt for them. Suitor had not flown since the Olympic Games opening ceremonies, a little over ten years earlier.

They contacted Suitor, who said he wasn't interested. He was soured by his last dealings with a rocket belt and he did not know Barker or the others. Barker continued pressing Suitor, and he sensed that Suitor did not believe the men had actually built a working belt. Barker finally asked if he could send Suitor the videotape of their rocket belt being test-fired in the shop. After viewing the tape, Suitor agreed to fly down and check it out.[6] Once again, Suitor was coming out of retirement to fly a rocket belt.

Suitor flew to Texas and visited Wright's shop where the American Rocketbelt Corporation did its work. Barker and the others had done a fine job of making a good-looking belt. Suitor noticed the bigger fuel tanks. He was concerned that the design changes would harm the belt's handling ability, though. Parts of it were painted bright red and much of the bare metal was shiny. Stanley would later say that Barker spent an inordinate amount of time polishing the device, like a teenager waxing his first car. He said Barker worshipped the belt. Suitor dubbed it Pretty Bird. Barker didn't like the name, but kept quiet. Suitor agreed to try flying it and then to train the others if the device worked.

Suitor strapped the rocket belt on and fired the motor. Although it made a lot of noise, it did not generate enough lift. Everyone knew what the problem was; it was the same problem Nelson Tyler had run into when building his belt. The belt looked right, but the throttle and motor assembly were not configured correctly. The interior parts were the key, and Barker had no idea what they looked like. They needed diagrams and designs for the inner workings of the throttle valve. Suitor told the men he knew a solution. He went and dug through some documents he had brought with him from New York. Among them was a diagram of the valve. Barker had the designs sent out and soon had a replacement valve for the rocket belt that would work.[7]

After fitting the new throttle valve into the Pretty Bird, Suitor agreed to try a tethered flight. They set up some tethers behind Wright's shop and Suitor put on the belt. He fired it up and lifted off the ground. He flew forward a few feet and then back. Although he was tethered, he knew the belt could fly. After running it for twenty seconds he touched down and told the men they had done it: they had built the second working civilian rocket belt. Suitor left for home.[8]

Then, the officers of the American Rocketbelt Corporation began fighting amongst themselves.[9] Stanley accused Barker of skimming money from the company, and Barker didn't think Stanley should ever fly the belt. Barker was younger and lighter than Stanley and told him that his weight was a problem. Stanley had been working on his weight and thought that Barker just wanted to be first to fly the Pretty Bird. Neither of them could agree on who would fly it first when Suitor returned to train them, and the arguments escalated.

While such an argument might seem trivial at first, it wasn't. Most of the pilots of the rocket belts needed thirty or more tethered flights before gaining the necessary proficiency to fly the belt untethered. Each practice flight expended a full load of fuel and was expensive. If Suitor was to train them both to fly the rocket belt, it could easily burn up sixty loads of fuel. The cost of the fuel alone could be anywhere from $100 to $500 per flight, depending on which sources would sell to them. And each flight required refueling, a time-consuming and somewhat dangerous procedure. Chances were that Suitor would train the first person to fly the belt and then leave that pilot to train the second. With the distrust between these two business partners growing, each suspected that the other would not follow through on training the second pilot.

By the fall of 1994, there were accusations of guns being pulled and actual fistfights between them. On one occasion, Wright had to take Barker to the hospital after a scuffle between Barker and Stanley at the shop had left Barker with a broken finger and some cuts and scratches. On another occasion, Barker attacked Stanley with a hammer, resulting in both men being arrested for assault.[10] The charges against Stanley were dismissed but Barker was convicted.

Barker convinced Wright to help him get the Pretty Bird away from Stanley's control. They met with an attorney and created a new company, Duratron Incorporated. They assigned the assets of the American Rocketbelt Corporation to the new company, even though they only owned half of ARC—and also drafted a lien they wanted to file against ARC.[11] They planned to argue that Wright was owed rent for the time ARC had used the shop and that this debt now gave them—Wright and Barker—the right to freeze the assets of ARC. And of course the only asset of the ARC was the rocket belt. Barker went to the shop and took possession of the Pretty Bird.[12] Stanley filed a lawsuit against the other two; among other things, he wanted the rocket belt returned. Barker and Wright acted as if no lawsuit had been filed and simply proceeded with their business plan without Stanley.

On January 21, 1995, the officers of Duratron Incorporated—Barker and Wright—brought Suitor back for more flight testing of the Pretty Bird. If Suitor determined the belt was safe to fly untethered, they could then discuss flight training for Barker and Stanley. Suitor was still concerned about the differences in the design of the Pretty Bird when compared to the Bell belts. To reacquaint himself with the belt, he tried it on a tether again when he first got to Texas. He did a few tentative hops, each time flying a little higher and for a little longer. Soon, he was convinced he could fly the belt safely without tethers. They agreed to take the belt out to a local airport the next day for a free flight.

On January 22, Barker, Wright, and a small group of their friends watched Suitor put on the rocket belt and fly. The group went wild. Suitor did two more flights, but the last flight ended in a rough landing. He was fine but the belt landed hard and was damaged. Suitor's concern about the enlarged tanks was proven correct: the changes Barker had made to the belt to give it a slightly longer flight time had made the belt unwieldy and harder to fly. And because the larger tanks were heavier, it used more fuel. The flights were not substantially longer than the flights had been in the old Tyler belt. Suitor later called the Pretty Bird "a fuel hog."[13] The longer flight time Barker had sought translated to a rocket belt that burned more fuel to get off the ground but only flew for twenty-three seconds, a gain of only two seconds.[14] And whenever someone asked, Barker claimed the belt could fly for thirty seconds.[15] Of course, Stanley had not been invited by Barker and Wright to see the belt fly.

In June 1995, the Pretty Bird was hired for an event. Duratron Incorporated had changed its name to the flashier American Flying Belt and had agreed to provide a rocket belt to fly over the Houston ship channel as part of a celebration for the Houston Rockets recently winning the NBA title.[16] And, despite his repeated attempts at retirement, Suitor found himself flying the rocket belt again. Barker and Wright received $10,000 for the flight, and paid Suitor $2,500 for his work. The flight lasted twenty-three seconds, although several sources gave the figure of twenty-eight seconds.[17] A twenty-eight-second flight would have been a remarkable improvement over the twenty-one-second flight limits of the old belts, and Barker and Wright wanted everyone to believe that their rocket belt was the best and longest-flying rocket belt available.

Otherwise the flight was uneventful as far as rocket belt flights go, until the landing. A local TV cameraman snuck into the landing area to get a shot of Suitor as he touched down. The cameraman got an earful of the rocket belt's exhaust and dropped his camera on the pavement. He later filed a lawsuit against Suitor but the mayor of Houston called the man's boss and asked them to drop the action, which they did.[18] After the flight, Barker loaded the rocket belt into a trailer and drove off with it. That was the last anyone ever saw of the Pretty Bird in public.[19]

Stanley saw the rocket belt flying on the news and waited for his day in court. Unfortunately, the case dragged on for a few years. After the successful flight, offers came in for more flights, but Barker had disappeared with the belt. It appeared that Barker was worried Stanley might take the belt back if the Pretty Bird was shown publicly. Wright also couldn't locate Barker, and Wright's audio business had gone defunct. He was in danger of losing his house and he turned to Stanley for help—even though Stanley was suing him. Wright asked Stanley how much the belt was worth to him. Would he pay Wright some money and drop him from the suit if Wright located it for Stanley? Stanley agreed to pay Wright $10,000 if the belt showed up in court at the next hearing. They reached an agreement—but Stanley never saw Wright alive again.

A few days later, Wright was found beaten to death in his own home. His body was so badly beaten that it took the police a few days to make a positive identification. The rocket belt never showed up in court.[20] No one was ever charged with Wright's murder, although many people believed there was an obvious suspect. The local sheriff's office named Barker as a suspect but said they had one other possible suspect too.[21]

The trial of Stanley's lawsuit was delayed several times but finally went before Judge John T. Wooldridge on July 26, 1999. It had been roughly four years since Barker had driven off with the rocket belt. Stanley showed up in court, but Barker did not. The court let Stanley put in his side of the case and over a period of a day and a half, he testified about how much money he would have been able to earn with the Pretty Bird. He also told the court how Barker had brutally attacked him with the hammer at Wright's shop. The judge granted a judgment in Stanley's favor in the amount of $10,242,813.76. More important, the judge awarded ownership of the Pretty Bird to Stanley.

Later, Barker told reporters that he did not know about the court date and that notices of the trial were sent to the wrong address—and that when they got there, people working for Stanley stole them.[22] Interestingly, he was aware of the court results within a day or two because he told reporters he had read about the trial in the newspapers. Even so, he never filed an appeal or took any actions to have the judgment—which he said was wrongfully entered—set aside.[23] Barker also didn't have any money to pay to Stanley and had no intention of giving up the belt. He refused to admit whether he had it or not. He was quoted in the newspaper as saying, "Even if I had it, I would smash it into a million . . . pieces with a road grader" before giving it to Stanley.[24] He then disappeared again. Stanley continued looking for him, but Barker kept on the move.

One day, Barker was contacted by a man he had gone sky diving with years earlier who now worked in Hollywood as a stuntman. He told Barker about a job opportunity working as a stuntman on a film in California. Broke and with few prospects on the horizon, Barker accepted the offer.[25] He traveled to California and met with the man. After some initial discussions at the man's house, the man pulled a gun out and told Barker he needed to tell him where the rocket belt was. Up to the moment the gun came out, Barker had no idea the film job wasn't real. The gunman and an accomplice tied Barker up and put him in a wooden box.[26] They questioned and beat him and left him in the box between questioning sessions.

One day they yanked him out of the box and handcuffed him to a chair. In walked Stanley with a woman who was a notary. Barker later testified that Stanley put some legal documents in front of him and ordered him to sign them at gunpoint so the notary could certify his signature. The documents purported to give legal title of the rocket belt to Stanley.[27] None of this made any sense, however. Stanley already had legal title to the rocket belt after the trial in Texas. Be that as it may, Barker signed the documents and was promptly put back into the box because he wouldn't tell Stanley where the belt was hidden.

After a little more than a week of beatings and being crammed into a box, Barker escaped and climbed out of a window in the house. He was in North Hollywood. He ran to a local restaurant and found a payphone. Soon the police descended on the kidnappers and arrested them, including Stanley, who appeared to be the mastermind of Barker's ordeal.

Stanley was tried in California for a host of crimes, one of which was kidnapping. After his conviction, Stanley was sentenced to a life term in prison. Part of the reason he got such a harsh sentence was that he refused to concede he had done anything wrong. He was clearly driven by his belief that he had the right to do something to get Barker to give up the rocket belt. Stanley's attorney summed up the situation after his client went to prison: "Unfortunately, what we have now, as I sit here today, is one man that's dead, an innocent man that's incarcerated, and you've got a con artist who's running free."[28] After he was sent to prison, Stanley finally realized he may have gone a bit too far. He said he was willing to admit he'd crossed a line or two. The court reduced the sentence to less than ten years after the prosecutor and Barker both asked the court to go easy on Stanley.

Barker was asked by ABC News if he still had the rocket belt. "If I had it, would I give it to Larry Stanley? No. No, I wouldn't." But did he know where it was? "I've answered that question as best as it will get answered, so move on to the next one." Pressed on whether the belt might turn up one day, he continued, "You know, you just never know. You never know."[29]

Back in Texas, Barker filed a suit against Stanley for civil damages arising from the kidnapping, but his suit was dismissed. However, Stanley's family had not given up on the huge judgment against Barker and they had not forgotten that he had the RB-2000. They complained to the judge who had overseen the original multimillion dollar lawsuit and now the judge was angry. After hearing Barker offer excuses and continually fail to follow the court's order to turn over the rocket belt, the judge threw Barker in jail for contempt. He spent six months in the local jail. The entire time, he refused to say where the belt was. [30]

Barker eventually got out of jail and moved back east. At one point, a writer contacted him and asked him about the Pretty Bird. Barker was coy and evasive, but it seemed clear he still knew where the belt was, or at least he wanted the writer to believe that he did.[31]

15

CIVILIAN ROCKET BELTS

While Barker, Stanley, and Wright were working on and fighting over the Pretty Bird, other people continued working in the field of rocket belts. Kinnie Gibson continued operating his Powerhouse Productions. After a while, the original Nelson Tyler belt Gibson used needed to be replaced, so he built a copy of it and eventually built another. He had a small fleet of three belts, counting the Tyler belt.[1] Since the Pretty Bird was nowhere to be found and the Bell belts had long since been grounded, Gibson's became the only actively flying rocket belt in the world.[2] Gibson hired more pilots to help with the flying duties, including Eric Scott and Dan Schlund.[3] Gibson didn't fly as often once he had trained Scott to fly the belt, although he did come back once or twice later for special events.

Schlund's training was not uneventful and was taking longer than expected.[4] After dozens of flights, Gibson began to wonder if the problem lay with the belt. They examined it carefully and determined that one of the exhaust tubes was a little bit longer than the other. It was a small difference that had not bothered Gibson or Scott because they were experienced pilots. Schlund had been trying to learn to fly with a belt that wasn't stable. Once that was fixed, the training went smoothly. On April 19, 2001, Schlund flew the Powerhouse rocket belt without tethers. The next day, the 40th anniversary of Hal Graham's first flight, Schlund got a phone call from Graham, who called to congratulate him.

The day after the phone call from Graham, on Schlund's third free flight, tragedy struck. Schlund came down too fast while he still had fuel in his tanks and hit the ground too hard. A man standing by to help him tried to right him and Schlund was thrown off balance. In a split second, the exhaust of the rocket belt blasted Schlund's leg. He woke up in the hospital with third-degree burns. He spent the next month there, undergoing three painful skin graft surgeries. Eventually Schlund made a full recovery, and he went on to make hundreds of flights for Powerhouse all over the world, in thirty-one countries on six continents.

Gibson's Powerhouse Productions traveled the world demonstrating the rocket belt and appeared in many television shows and movies. They also appeared at countless venues such as sporting events and air shows. On its website Powerhouse listed a variety of appearances made. One of the appearances listed was the 1984 Olympics, when Bill Suitor flew the Nelson Tyler belt.[5]

Not all appearances were credited to Gibson's company, but that was part of its business plan. Musician Sean "Diddy" Combs gained a huge amount of media attention when he flew into a press conference for the 2005 MTV Video Music Awards via rocket belt. At least, that is how it appeared to people who saw the cleverly edited video footage of a rocket belt pilot in a white suit flying through the air and then saw Combs on the ground, carrying the rocket belt pilot's helmet and wearing a white suit. People with access to the Internet could see on the Powerhouse Productions website that the stunt they had just seen was one anyone could buy. "Ideas in how best to utilize the Rocketman" included "Execute a switch-out with your actor, company president, or CEO."[6] People who want to watch the video of the flight can find it on YouTube, watermarked with a URL that directs viewers to the website for Powerhouse Productions.[7] The man flying the rocket belt was Schlund, making his first public flight for the company.[8]

In 2004, Gibson filed a trademark application for the name "ROCKETBELT," saying that he had been the first to use it in 1981. In 2008, he filed an application to register the name "ROCKETMAN" as a trademark, likewise claiming he had first used the name commercially in 1981. The trademark office asked Gibson to provide some evidence of his use of the word "rocketbelt." In response, he sent the US Patent and Trademark Office a stack of advertising fliers and literature outlining the business,

Powerhouse Productions, Inc., and appearances made by "The Rocketman."[9] In the cover letter, his wife wrote,

> We (my husband & I and our Company "Powerhouse Productions Inc.") has used Rocketbelt since the early 80's to identify our machine/s. The word really exploded when we did the 1984 Olympics and we had another rush when we posted ourselves on the internet. . . . There are many that would like to ride on the coat tails of what we have worked hard to create. Can you please help us to secure this word?[10]

Regarding the 1984 Olympics, a Rocketman brochure accompanied the correspondence that said:

> The Rocketbelt is the most spectacular instrument ever designed for super crowd pleasing entertainment and product endorsement. There is nothing like it in the world!
>
> . . .
>
> Amid the spectacle, one event out shadowed them all. . . . Rocketman thrilled billions worldwide when he flew into the Los Angeles Stadium with just a rocketpack on his back.[11]

Of course, Bill Suitor was the one who flew the belt at the 1984 Olympics. The documents also list other places the Rocketman was said to have flown, including the fair in Adelaide and the Knoxville world's fair.

Many others in the rocket belt community believed that the claim about the name "Rocketman" was wrong. A pioneer in the field of rocket technology named Ky Michaelson had been building and racing rocket-powered motorcycles and cars for decades. He held racing records for rocket-powered devices in wildly divergent categories, from a rocket-powered snowmobile to a rocket-powered car. A magazine summed up his career by calling him "The Edison of crazy."[12] In 1964 or 1965, a drag strip announcer introduced him to an audience as "The Rocketman," a name he began using regularly afterward. Most of the vehicles that he raced were powered by giant hydrogen peroxide motors. A rocket car he built was the first dragster to officially clock three hundred miles per hour on an NHRA drag strip; later, it was the first car to break into the "fours"—an elapsed time of less than 5.0 seconds in a standing-start quarter mile.[13] Another car he built ran over four hundred miles per hour in a 3.22 second quarter mile, although some sources note that it wasn't on a sanctioned NHRA drag strip. The driver of that car, Kitty O'Neil, later clocked in at 392 miles per hour on an official NHRA track with the Michaelson-built rocket car.[14] The fastest Michaelson had personally ever run was *only* 390 miles per hour, also in a rocket-powered car.[15] He was well known in the rocket vehicle community and sold rocket motors to many other drivers. The name *Rocketman* was so ingrained in Michaelson's identity that his son's middle name is *Rocketman*.

The US Patent and Trademark Office registered Gibson's trademark on October 4, 2005, for *Rocketbelt*, and on February 24, 2009, for *Rocketman*.[16] Trademark attorneys are quick to point out that the simple registration of a trade or service mark is not an indication by the government that the registrant was, indeed, the first person to have used the mark. It is merely an indication that they were the first to apply to register it.

Ky Michaelson said someone called and told him he could no longer call himself the *Rocketman* now that the name had been trademarked by someone else. "Who are you trying to kid? That's my name," was his response. He kept using it. Likewise, his son kept using it as his middle name.

Regardless of what they were called, more people began building their own individual lift device (ILD) belts. People began sharing information on the Internet, and whole websites were devoted to rocket belts, some of which included boards for enthusiasts to discuss and trade their ideas. Some website operators began selling parts and materials to build rocket belts. Somehow copies of the plans for the notorious throttle valve from National Water Lift found their way onto the Internet. Some people suspected that the plans had resurfaced after National Water Lift had made the replacement valves for Kinnie Gibson in 1993. One website simply posted the plans for anyone to download.

At least one rocket belt builder received a letter from a law firm in Texas claiming to represent Gibson's company. The letter warned its recipient to remove the valve drawings from the Internet and demanded to know where the plans had been found. According to the letter, Powerhouse "owns the manufacturing and proprietary rights concerning the Rocketbelt and the throttle valves used in same. Powerhouse has exclusively owned said rights for several years."[17]

The letter was interesting for several reasons. First, it did not explain how Gibson came into ownership of the manufacturing and proprietary rights of the rocket belt. Wendell Moore was granted a patent for the rocket belt in 1962; presumably, the patent had expired many years earlier. In fact, copies of the patent had been sold out of the back of *Popular Science* for years before this. Interestingly, the valve does not appear to have ever been patented. In the belts, built more than forty years earlier, the valve bears no patent information whatsoever. The letter offered no description of the rights held by Gibson, or how he could have come into them. National Water Lift was bought by another corporation in the 1990s and eventually fell under the umbrella of Parker Hannifin, a multinational aerospace corporation headquartered in Cleveland. In 2012, a spokesperson for Parker confirmed that the rights to the valve design were not sold to anyone.[18] It would appear that the rights to the rocket belt and its various components are within the public domain at this time.

Typical of the civilians who set out to build a belt for no other reason than a fascination with the idea was a thirty-six-year-old man from New Jersey named Gerard Martowlis. Around 1990, Martowlis set out on a project that would end up devouring his few spare moments between raising a couple of children as a single father and holding down a full-time job. In the corner of his basement, the rocket belt slowly took shape over seventeen years. Martowlis was not an aerospace engineer. Still, he was handy and worked in a technical industry that made him familiar with tanks and valves.[19]

Martowlis began researching and designing the belt on his own, long before the Internet was able to answer any of his questions. While he was mechanically inclined, he farmed the machine work out to others. The hard anodized aluminum parts for the throttle valve, in particular, were a stumbling block. He found the plans for the valve but then found out how hard it was to get someone to actually make one. Some shops in New Jersey that could make the parts balked when they found out what he was building. No one wanted to make parts for his rocket belt, fearful he might fall out of sky one day and sue them. After being rejected by a few shops, he found one that would do it when he told them the valve was for a high-pressure sprayer, like a power washer. They machined the valve parts for him, never asking why his sprayer needed such precision craftsmanship. The machinist told him afterward that if he were to hold the plunger from the valve in his hand for a bit, the warmth from his hand would cause the piece to expand more than the clearances inside the valve, and not to worry if the pieces didn't seem to fit together. When they were both at the same temperature, they fit just fine.[20] Martowlis also found a welder who could join the various aluminum pieces of the belt to his satisfaction.

Martowlis stated, "I didn't have the information available in the 1990s that I do now. I did most of this stuff off the cuff, without anyone's help." Martowlis studied rocket motors the old-fashioned way, before Google came along to make everyone's life simpler. He decided to beef up the motor for safety reasons, and then cut weight from the harness and corset, again for safety reasons. After getting enough of his belt built to where he could test the motor, Martowlis went looking for hydrogen peroxide. By then the Internet was available to help him. He found one of the most knowledgeable people in the world on the topic of rocket belts: Peter Gijsberts of the Netherlands.

Gijsberts had always been fascinated by rocket belts. Born in 1962, he lived in the town of Nuenen, famous for Vincent Van Gogh's early years. Gijsberts saw vintage footage of Suitor and Courter flying over a field and through some woods in a rare film where the two pilots wore cameras. The film not only showed what the men looked like in flight, but also what they saw. Gijsberts was mesmerized. Later, searching for the devices on the Internet, he wasn't sure what to call them. He didn't know the devices were called rocket belts, and English was not his first language. He queried search engines with terms like "flying man." He eventually found what he was looking for, but decided it should have been easier. The information belonged in one place.

It was a testament to how widely they had been publicized that a man so far away from Buffalo, New York, would become one of the biggest promoters of rocket belts. Gijsberts launched a website, www.rocketbelt.nl, that became a focal point for rocket belt enthusiasts around the globe. He researched the history, collected documents and photos, and tracked down people who had worked on the program. Soon he became the go-to guy for rocket belt information, no matter how trivial or obscure. He also organized a discussion group on Yahoo!

It was in this group that Martowlis asked about locating hydrogen peroxide. At first he was coy. He had been keeping his rocket belt project relatively low key; only friends and family knew of his work. Gijsberts and the others were curious. Why did he need hydrogen peroxide? Martowlis decided there was no further need for secrecy. He told the group he had a completed rocket belt and was ready to fire it up. All he needed was fuel. The group buzzed with excitement. Many of them had discussed building the belts and some had even started. Now a stranger had completed one!

After the buzzing tapered off, someone told him to contact a man named Erik Bengtsson who owned a company in Sweden called Peroxide Propulsion. Martowlis ordered the fuel for his belt. After consultation with other members of the group, which now included some actual rocket belt pilots such as Bill Suitor, Martowlis took his belt into his backyard for testing. He strapped the belt and its stand to a tree a foot thick so he could stand on the other side of the trunk. If something blew up, he hoped the tree trunk would shield him from the shrapnel. "To tell you the truth, I had butterflies in my stomach. The adrenaline was definitely pumping at the time. I had no idea. I had never fired one of these things before. It's a scary machine." He fired his belt and the motor ran fine. He filmed the test and posted it so the other members of the community could study his results. His rocket would run for twenty-five seconds and he is certain that his belt is one of the lightest built to date. Empty, his belt weighed "around sixty pounds."

From there, Martowlis took the plunge and strapped the machine to his back. He started with small fuel loads and took bunny hops to get used to the feeling of being lifted by the belt as advised by rocket belt pilots who frequented Gijsberts's websites. Even so, he hit some rough spots. At one point he fell over while the belt was running and the machine dragged him across his lawn, cutting small trenches in the ground for ten feet. His original valve was a bit different from the National Water Lift design. After more tests, he determined that his throttle valve needed reworking. The throttle did not allow him enough control; it seemed to either be full throttle or nothing. In between it wasn't specific enough. Martowlis decided to tear the belt apart and rework it.

In 2012, he was still working on refining the belt. When he is done, he hopes to take flight. He is already aware of places where he can make paid appearances. The project had started in 1990, and Martowlis is not in a hurry. He didn't keep track of his expenses along the way but he guesses his work could be replicated for $20,000 to $30,000. He is quick to point out that the fuel costs sneak up on many of the rocket belt pilots. The fuel is expensive and burns at about six seconds per gallon; a person can burn through quite a bit of money very quickly while playing with hydrogen peroxide.

Meanwhile, he stays in touch with the other rocket belt builders, sharing expertise. When one professional rocket belt team flew nearby, Martowlis loaned them a few gallons of fuel to save them the trouble of having to transport the stuff cross-country. He also gives advice and, like the others in the small community, hopes others who follow him don't do anything stupid. "If anything were to happen, God forbid, if they were to hurt a bystander or burn someone," things would get very difficult for these men to continue their hobby. The rocket belt builders don't mind attention; they just don't want to be viewed as a danger to the community.[21]

Whenever he gets the chance, Martowlis watches flights by other pilots. He has seen Eric Scott fly the rocket belt several times. One day he saw Scott land, take off the belt, and pull out a cigarette. Needing a light, he put the cigarette in his mouth and leaned over to the gas generator—the motor—of the rocket belt and touched the tip of his cigarette to light it.

The people working on today's rocket belts always want to extend the twenty-one-second flight limit of the old Bell belts. Using modern technology, one would think it would be possible. Perhaps using lighter, space-age materials and larger fuel tanks would help? People familiar with the Pretty Bird knew that simply enlarging the fuel tanks didn't add enough performance to make up for the gain in weight or the diminished control resulting from the bulkier belt. Still, little by little, the flight times have inched upward. Materials like carbon fiber and Kevlar are used in the modern belts, shaving weight and adding strength. By 1995, the state-of-the-art civilian rocket belts, such as Gibson's two newer models, had reached thirty-second flight times, according to the builders.[22]

Even with the blueprints available on the Internet, the valve remains a stumbling block for many would-be rocket belt builders. The gas generator also is not something someone could just buy at the local hardware store, though many of the other parts of the belt are off-the-shelf hardware anyone can buy: tanks, tubes, belts, nozzles. Some people have put belts together that resemble the original Bell Rocket Belt but cannot get off the ground. The valve and the gas generator need to be machined precisely, in a manner that might take hours of work by a skilled machinist with a good machine shop. Hiring a machinist to do the work might cost several thousand dollars. Someone who is a machinist and willing to invest his own time could make the valve and the gas generator for a very small sum: just the cost of a few chunks of metal. Ky Michaelson has made a few rocket engines and has said making a good throttle valve and gas generator remains a tricky proposition for anyone. "There are a lot of very small holes in that thing that are very hard to machine."[23]

In 1997, another company joined the rocket belt fray. Carmelo "Nino" Amarena graduated from the Buenos Aires University of Engineering with a master's degree in electro-mechanical engineering. He was partly inspired by his dreams of flight, fueled by cartoons and children's shows he'd seen growing up in Argentina. Many of the shows were Japanese imports that featured rocket belts and jet packs. Shortly after he graduated from college he moved to California, where he worked on a variety of projects for racing teams and for NASA. While lamenting the traffic during his commute one day, he wondered about alternatives to the bumper-to-bumper grind he found himself in each morning. What if someone made the jet packs he had seen in the cartoons as a child? He found some investors and Thunderbolt Aerosystems, Inc. was founded with the goal of manufacturing rocket belts for sale to the public. Amarena designed

Science fiction artist Frank Paul drew one of the earliest "jet packs" when he illustrated "The Skylark of Space" for *Amazing Stories* in 1928. *Reprinted with the acknowledgement of the Frank R. Paul Estate.*

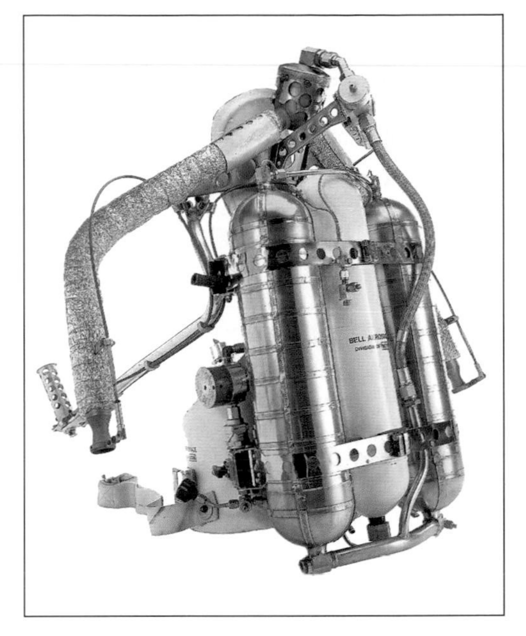

Bell assembled five rocket belts, four of which survive today. Bell Rocket Belt #2 is in the collection of the Smithsonian National Air and Space Museum. *Photo by Carolyn Russo, National Air and Space Museum, Smithsonian Institution (SI 2007-15054).*

Bill Suitor flies above the New York state fair in 1969, shortly before his first retirement from flying rocket belts. This was also the last public appearance of the Bell rocket belt flying team. *Photo courtesy of William Suitor.*

Governor Nelson Rockefellor sits on a curb and watches Bill Suitor fly the rocket belt over the New York state fair in 1969. *Photo courtesy of William Suitor.*

The notorious throttle valve of the rocket belt, manufactured by National Water Lift of Kalamazoo, Michigan. The valve would prove to be the stumbling block for many later rocket belt aficionados. *Photograph © Evelyn Ihrke 2012.*

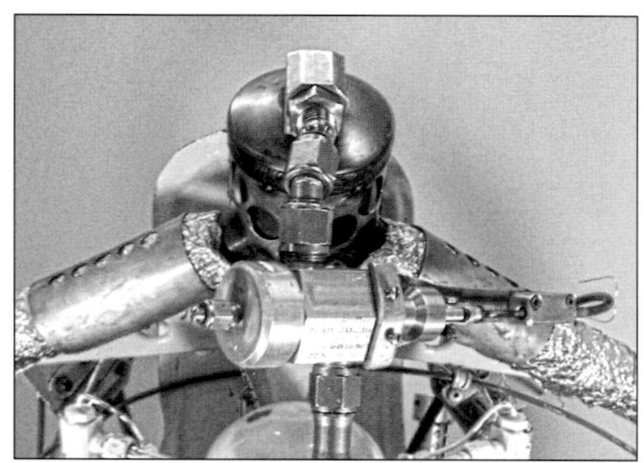

The gas generator of the rocket belt, where hydrogen peroxide interacted with a catalyst bed and burst into steam in fractions of a second. *Photograph © Evelyn Ihrke 2012.*

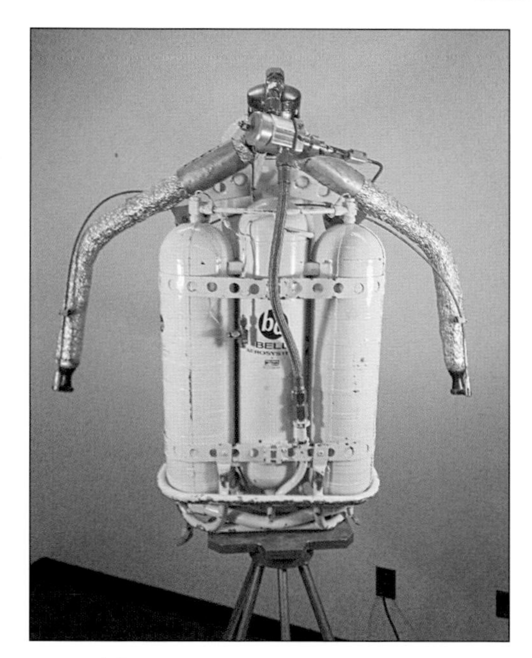

One of the surviving rocket belts is held by the Industrial & Systems Engineering Department in the School of Engineering and Applied Sciences at the University at Buffalo. *Photograph © Evelyn Ihrke 2012.*

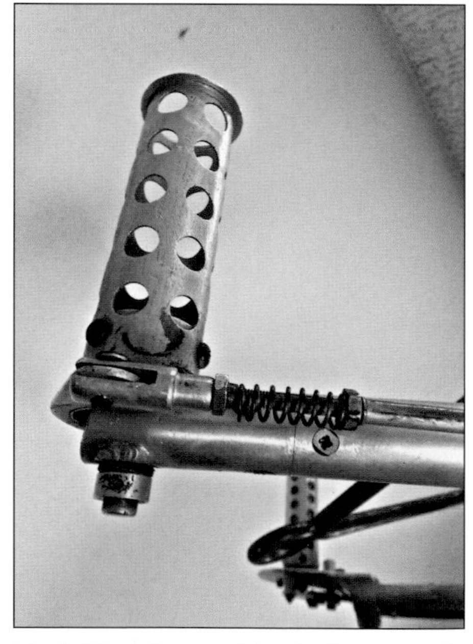

The left hand control for the "jetavators" allowed the rocket belt pilot to control yaw in flight. *Photograph © Evelyn Ihrke 2012.*

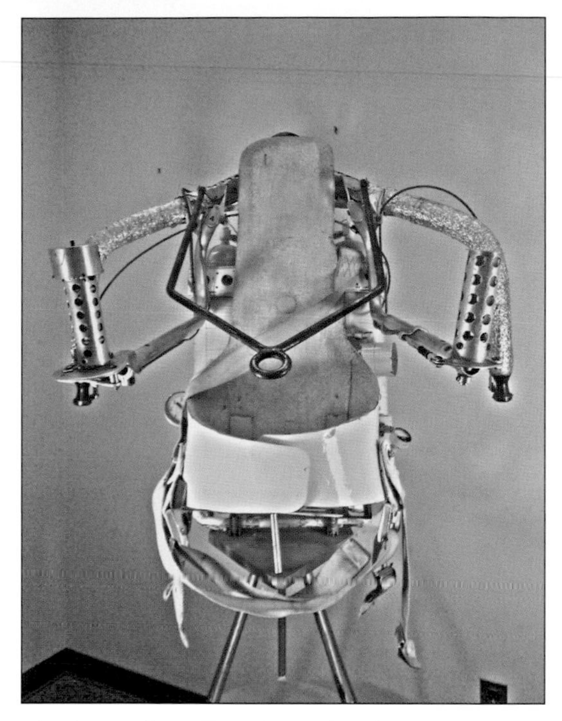

The front of a Bell Rocket Belt. *Photograph © Evelyn Ihrke 2012.*

Bill Suitor successfully flies the "Pretty Bird" rocket belt which later became the center of controversy. One of its owners would be murdered, one would spend time in prison and the third would be kidnapped after the belt disappeared. To date, no one knows the whereabouts of this device. *Photo courtesy of William Suitor.*

Williams WR-19 Turbofan engine, serial number 3. An engine identical to this one powered the Jet Belt. Only 12 inches wide and 22 inches long, it was a marvel of engineering. It put out over 400 pounds of thrust and weighed less than 60 pounds. This engine is in the collection of the Smithsonian. *National Air and Space Museum, Smithsonian Institution (SI 2011-00999).*

A Williams F107-WR-101 on display at the National Museum of the U.S Air Force in Dayton. In this configuration, it weighs less than 150 pounds and generates 600 pounds of thrust. This engine is used to power the Air Launched Cruise Missile. *Photo by author.*

The WASP II stunned observers with its ability to hover, fly forward and backwards, and duck behind trees. Here it is in flight, with Bob Courter at the controls. *Courtesy of Mark Voss.*

Peter Kedzierski, Hal Graham, Bill Suitor, Peter Gijsberts at the first Rocket Belt Convention in Niagara, 2006. *Photo courtesy of Peter Gijsberts.*

A WASP II, painted as an X-Jet, suspended from the ceiling of the National Museum of the US Air Force in Dayton. *Photo by author.*

Rear view of the WASP II. *Photo by author.*

Side view of the WASP II shows one of the two air intakes, covered by a screen. The hole below the duct is for attaching tethers or the gimbal during training. *Photo by author.*

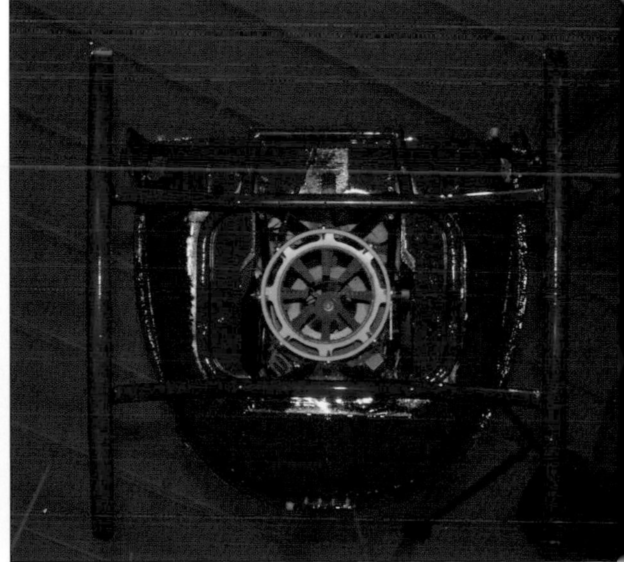

The underside of the WASP II, showing the exhaust and the landing skids. The vanes within the exhaust duct turned in conjunction with the yaw control. *Photo by author.*

Isabel Lozano became the first woman to fly with a rocket belt in 2006, wearing a custom-made belt designed and built by her father, Juan Lozano. *Photo courtesy of Juan Lozano.*

Juan Lozano flying a rocket belt he built himself. *Photo courtesy of Juan Lozano.*

Nick Macomber flies the Go Fast! rocket belt over the California coastline. *Photo by Jeff Folk, courtesy of Go Fast!*

A Springtail flies while tethered, with Robert Bulaga at the controls. Note that the pilot has taken his hand off of the left control to give the "Thumbs up" sign. *Photo courtesy of Trek Aerospace, Inc.*

Mike Moshier, founder of Millennium Jet, at the controls of the SoloTrek. *Photo courtesy of Trek Aerospace, Inc.*

"Jetman" Yves Rossy. A Swiss pilot with fighter pilot and commercial flying experience, he has also flown his jet-powered wing, using nothing but his body movements to steer. *Photo © André Bernet/ Jetman.com.*

A Jet Lev pilot hovers, held in place by water blasted from a jet pack-like device on his back. The unit looks and operates much like the rocket and jet belts but remains tethered during flight and can only operate over water. *Photo courtesy of Jet Lev.*

Rossy's first successful wing used two small jet engines originally marketed to radio-control hobbyists. *Photo courtesy of jetman.com.*

In order to fly faster, Rossy added two more engines to his wing. With this wing, his flight speeds often approach 200 MPH. *Photo courtesy of jetman.com.*

Rossy has made many spectacular flights, including this one over the Grand Canyon. He has also flown over the Alps, Rio de Janeiro, and across the English Channel. *Photo courtesy of jetman.com.*

and built a model and asked Bill Suitor for help in testing it. Suitor, who had realized he would probably never be able to retire completely from the rocket belt game, agreed. The company soon advertised Suitor as their director of training.[24]

Amarena, like many of the rocket scientists before him, could not resist the temptation to fly himself. He strapped on his ThunderPack and flew, finally fulfilling his childhood dream. In 2003, Thunderbolt Aerosystems said it had sold a ThunderPack to a customer "in the far east" who planned to use it for "rescue purposes." The ThunderPack R1-G2 was a bit different from its forerunners. By adding kerosene, diesel, or methanol to the hydrogen peroxide, flight times had been increased to a claimed thirty seconds. Thunderbolt offered these belts for $125,000 for a time and then announced a new model. After two years of testing, they announced the R2-G2, a belt that could be configured for the hydrogen peroxide–only setups of old, or as a dual-fuel burner like their R1-G2. In a dual-fuel configuration, they claimed flight times exceeding a minute—but many in the rocket belt community were skeptical. Long gone were the days when someone could make extravagant claims about rocket belt performance without being questioned. Without public demonstrations or videos posted to YouTube, the claims would be largely ignored.

The appeal of the rocket belt was global. A Swiss engineer named Arnold Neracher began working on hydrogen peroxide rockets in 2000 or 2001. To him, it was just an engineering problem to be solved and soon he had a rocket belt. As many other engineers would do, he put his motors on other vehicles such as a bicycle and a car. Seeing the scarcity of the fuel, he also built his own machine to refine the hydrogen peroxide himself. Neracher was one of the few rocket belt builders who resisted the urge to fly his belt. He enlisted the help of a Swiss pilot named Yves Rossy to test his belt for him.[25]

Sources of high-grade hydrogen peroxide were hard to find, but the rocket fuel was available. Eric Bengtsson, the chemical engineer in Sweden who had helped Martowlis get fuel, had founded Peroxide Propulsion in 2003 in the town of Gunnilse, not far from Gothenburg. He had been working with the chemical for some time and recognized the specialized market as an opportunity. Soon, he was supplying the liquid to rocket belt enthusiasts and others.[26] His hydrogen peroxide could be shipped worldwide. One man in the United States paid $12,000 to have thirty-six gallons shipped over, which breaks down to $333 per gallon.[27]

In 2003, a man in Denver named Troy Widgery and two of his friends formed a company called Jet Pack International LLC with the goal of building a rocket belt. Widgery owned a company called Go Fast Sports, which marketed an energy drink, and he had always been fascinated by rocket belts. He thought they might even be helpful in marketing his product. After all, what else does a rocket belt do besides go fast? Widgery had met Ky Michaelson, the Rocketman, a few years earlier, when Michaelson was looking to become the first civilian to launch a rocket into space. Widgery lent financial support to Michaelson's successful efforts and soon Michaelson was lending his hydrogen peroxide expertise to the Jet Pack International project with Go Fast.[28] Widgery and company also connected with a man named Jeremy McGrane, who had spent much of his adult life studying the Bell Rocket Belts. A self-taught machinist, he had set out to build a belt himself. He sold his first completed belt to Widgery.

Ky Michaelson also had some helpful documents in his possession. When Bell closed the facility where the rocket belt work had been done, much of the material from the program was simply discarded. Bell paid someone to dispose of it, and that person wondered if any of it had value. A few phone calls later, Michaelson had agreed to buy anything and everything to do with hydrogen peroxide technology that was otherwise going to a landfill, sight unseen. Michaelson agreed to give the man $1,000 for everything he could box up and ship to Minnesota. After getting a pile of boxes, he got a phone call from the man saying there was still a large amount of material left that he hadn't been able to box. For another $1,000, Michaelson could have the rest but he'd have to come and get it. Michaelson made a road trip to New York and brought back a carload of boxes. As he dug through them he found papers, rocket parts, and blueprints.[29]

Other than the valve, the rest of the rocket belt is not that complicated to build, according to Michaelson. "You could build a belt for less than [the cost of] a Harley Davidson." The best catalyst bed is made of nickel, silver, and samarium nitrate, which isn't even all that difficult for the garage-based rocket builder to come up with, although some builders say that silver works fine by itself. "If you're a halfway decent machinist and fabricator, you can build that belt easily."[30]

The McGrane rocket belt bought by Widgery was built along the lines of the Bell Rocket Belt, but substituted modern materials wherever possible. When it was done, the Go Fast Jet Pack weighed 135 pounds and could stay aloft for thirty-three seconds. Interestingly, that belt was *heavier* than the Bell belt, which weighed only 110 pounds fully fueled.[31] The Go Fast Jet Pack had taken three years and nearly a million dollars to develop.

Around 2005, Eric Scott met Widgery. Scott had worked with Powerhouse Productions and Kinnie Gibson. Scott had not only flown Gibson's rocket belt, he'd also done stunt work on *Walker, Texas Ranger*, where he'd routinely get beaten up by Chuck Norris, or have himself set on fire. Somehow, flying a rocket belt seemed like part of the routine. Scott told a reporter, "It's amazing. The feeling of flight is really hard to explain. To defy gravity. It's like jumping off the ground and not having to come down right away." By 2007, Scott was forty-four years old and had already made more than six hundred flights in the rocket belts.[32] Widgery hired Scott to fly the Go Fast Jet Pack.

As more people began flying the hydrogen peroxide–powered rocket belts, the odds of an accident increased. The only accidents outside of Bell's program had been Schlund's burn and Gibson's broken knee, and those weren't accidents with the fuel. There had been other occasional scraped knees and rough landings. The people handling the fuel, however, ran into their own problems. In August 2005, a truck carrying hydrogen peroxide manufactured by Erik Bengtsson in Sweden was traveling the M25 highway outside of London when something onboard caught fire. "Motorists described hearing a series of explosions and seeing items blown across all ten lanes." The roadway itself was damaged by the heat and the truck was burned beyond repair, but only the trucker was injured, and he was lucky to not have been killed. Drivers were stranded for hours while trying to get to nearby Heathrow Airport.[33]

It was a scary moment for Londoners, who were a bit on edge after the events of July 7 of that year. That day, terrorists detonated four bombs in and around London, on subway trains and on a double-decker bus, which was "ripped open like a can of sardines."[34] The memory of that event, which had killed fifty-six people, was still clear in people's minds. What fewer people knew was that the bombers had used hydrogen peroxide. It turned out that the fuel that propelled Graham, Suitor, and the others into the sky was also an ideal ingredient for homemade bombs. And it was frighteningly easy to acquire. After an inquest was held in London, the coroner criticized its availability. "So you get cross-examined by the chemist if you want to buy too many aspirin, but you can buy as much hydrogen peroxide on the market [as you want]." The bomb makers had simply walked into pharmacies and bought entire stocks until they had as much as they needed. They took the liquid back to a flat in London where they refined it to a higher concentration and then packaged it into portable bombs made out of items from the local hardware. The finished products weighed only a little over twenty pounds. The hydrogen peroxide seemed so innocuous that the bombers "made no attempt to disguise their work."[35]

At the time of the London subway bombings, other terrorists were busy distilling hydrogen peroxide. Police in London broke up and arrested a ring of would-be bombers who wanted to replicate the events of July 7 and had gone so far as to manufacture half a dozen hydrogen peroxide bombs. The men were arrested before any of their bombs detonated but police found a bomb factory much like the one of the July 7 bombers.[36]

The activities of more would-be terrorists in Britain would soon affect air travelers worldwide. British law enforcement conducted surveillance of a group of seven men who were preparing bombs and in August 2006 they arrested the men before they acted. When descriptions of the failed bomb plot were made public, it caused dramatic changes in air travel. The men had been planning on using hydrogen peroxide bombs to blow up transatlantic flights to cities in North America. Authorities believe the men would easily have managed to get their bombs through security because they were primarily just containers of liquid. They were going to smuggle the hydrogen peroxide aboard the plane inside soft drink bottles.[37] An x-ray of the bottles would not have tipped anyone off to what was inside. Hydrogen peroxide looks no different than water or 7-Up.

These bombs were much smaller than the subway bombs but experts were convinced they would have succeeded.[38] In the aftermath of this incident, the US Transportation Security Administration passed new rules regarding liquids on airplanes. Travelers would be restricted to no more than three ounces of liquid in their carry-on luggage, much to the annoyance of parents traveling with small children and travelers who packed regular-sized bottles of shampoo or hairspray. The TSA was coy about the rationale behind the new rule. On its website, the TSA said, "TSA and our security partners conducted extensive explosives testing since August 10, 2006, and determined that liquids, aerosols and gels, in limited quantities, are safe to bring aboard an aircraft."[39]

Frustrated travelers were confused by the way the new rule was presented by the TSA. It sounded as if the agency was asserting that typical carry-on luggage had been dangerous and it took extensive explosives testing to determine that three ounces of hair gel wouldn't explode and drop a 747 from the sky. What the TSA wasn't saying, which would have made more sense, was that it had been determined to be too easy to smuggle a hydrogen peroxide bomb onto an airplane under the old rules.

Meanwhile, more rocket belt enthusiasts were building and flying their own belts. Juan Manuel Lozano founded his company, Tecnologia Aerospacial Mexicana, in Mexico. Lozano lived in Cuernavaca, and in the early 1960s, a young Lozano had attended a technological exposition where space-age technology was displayed. Along with space capsules and astronaut suits, a Bell Rocket Belt flew two times each day in the polo grounds next to the auditorium.[40] The youngster was hooked and told himself he would build one someday.

First, his research revealed a fuel shortage. Hydrogen peroxide was becoming increasingly difficult to obtain commercially. Lozano was inventive and decided to cut out the middlemen. How hard could it be to make? He would not only make his own fuel but he would design and build a machine to automate the process. He did that first while studying and designing his rocket belt. He realized he could sell the hydrogen peroxide machines and turn a profit, making money he could reinvest in his project. It turned out that hospitals were prospective consumers of the hydrogen peroxide refining machines, and at least one machine was sold to another company flying a rocket belt.[41] Lozano sold twenty of the machines for $15,000 each.[42] Once he was assured an unlimited supply of refined hydrogen peroxide, he set out to build belts. In later years, rocket belt owners reported the price of Lozano's machines had crept up to $20,000.

Lozano eventually built nine rocket belts, a Pogo-style device, and put rocket-powered engines into motorcycles and go-karts. He estimated the projects cost him more than half a million dollars over thirty years.[43] Lozano, like many of the rocket belt developers, built his rocket belt hoping he could fly it himself. He first flew it on April 9, 2006.

In the rocket belt community Lozano is perhaps best known for one other thing: on August 11, 2006, his twenty-nine-year-old daughter became the first woman ever to fly a rocket belt. Lozano custom-built a pink harness for Isabel Lozano and even designed a set of tanks matched to her size. In her second flight she reached an altitude of eighteen feet. Video footage of her flights was shown on television and on the Internet. Soon, millions of people had witnessed the flight of the first "Rocketwoman."

16

THE MODERN ROCKET BELT IN THE PUBLIC EYE

Peter Gijsberts met Kathleen Lennon Clough, the daughter of Bell's official photographer, through his website and they decided to hold a rocket belt convention in Buffalo, near the old Bell facilities. They secured the Niagara Aerospace Museum for a weekend and began sending out announcements. They wondered what kind of reaction they would get. The press release announced the "First International Rocketbelt Convention: Where the Past Meets the Present," to be held in Niagara Falls, New York, on September 23 and 24, 2006.[1]

Word spread on the Internet and soon people from around the world said they would attend. "We pulled together [the] convention in less than five months . . . but we had enthusiasts coming in from Australia, Sweden, Spain, all over."[2] Gijsberts flew over from the Netherlands and hosted many of the people he had been communicating with and wanting to meet. Former Bell employees were admitted for free. In total, 175 people showed up. Hal Graham—now seventy-two years old—flew up in his own plane and took the stage to speak about his experiences flying the rocket belt. He wore his rocket belt flight suit for his talk. It was tattered but still fit, more or less. He then pulled out a ukulele and sang a song he had composed entitled "My Rocketbelt Daze." Peter Kedzierski showed up, as did Bill Suitor.

Erik Bengtsson came over from Sweden and told the audience about his business, Peroxide Propulsion, which sold hydrogen peroxide to many of the rocket belts then flying. Some people in the audience whispered about the truckload of his fuel that had exploded on the highway near London a year earlier. John Spencer, the head test pilot from Bell, spoke. Doug Malewicki told of his involvement with the Pretty Bird. He had designed the catalyst pack and some nozzles for the rocket belt before all the drama and tragedy struck. He would later say that the design work on these parts was something "any rocket propulsion college student could have done."[3] Nancy Wright, Joe's sister, spoke about her brother's unsolved murder and the story of the Pretty Bird.

The new generation of rocket belt builders was well represented at the convention as well; one of its stated purposes was to give these builders an opportunity to meet the first generation. Nino Amarena was there and, keeping with the relatively low-budget atmosphere of the convention, he helped Gijsberts and Clough with the technical aspects of the auditorium such as lighting and sound. Amarena's master's degree in electro-mechanical engineering overqualified him for thc task, but he was happy to help. Troy Widgery and Eric Scott were there and made the loudest demonstration when Scott flew the Go Fast Jet Pack just outside of the museum for the grand finale.

It was an interesting moment in history: the Go Fast belt was visually a carbon copy of the Bell belts that had been flown by many of the men in the audience decades earlier. John Spencer, the test pilot who had flown the Bell belt a hundred times, was appalled when he saw how the Go Fast crew flippantly handled the hydrogen peroxide. "To me, it was scary."[4] Of course, Spencer's men had been handling the fuel decades earlier, when much of what they were doing was still considered experimental. The knowledge base and experiences of the Bell men had proven the technology was safe.

Carolyn Baumet, Wendell Moore's daughter, attended and told stories of Moore's involvement with Bell. At one point she unveiled a drawing her father had made, imagining what the rocket belt would look like. She announced she was donating it to the museum.

Hugh Neeson, the former vice president of Bell, and Bob Roach, the project engineer, both spoke about the Bell program. The audience was peppered with other former Bell employees and children and relatives of people who worked with the belts. And there were quite a few people there who just wanted to build one of their own. Some had begun work and others were still gathering information and looking for tips. A writer named Mac Montandon visited the conference on a personal odyssey to fly in a jet pack and gave the dreamers and hobbyists equal time when he interviewed attendees, trying to get a feel for what drove these guys to build their rocket belts.

A couple of the original rocket belts were on display, and some modern ones as well. Suitor brought the flight suit he wore in his famous appearance over the opening ceremony at the 1984 Olympics. If anyone wondered about the popularity of the rocket belt before the convention, the enthusiasm at the conference extinguished those fears. The media had not forgotten the rocket belt either. PBS and the History Channel sent film crews and Gijsberts counted five television reporters. *Slate* magazine had a writer there, and so did some publications from Europe.[5] Canada's History Channel and websites such as Space.com and BoingBoing.net covered the convention.[6]

Suitor, in particular, was often the center of attention, lecturing and answering questions and giving press interviews. He described his worst ride to a writer from *Slate*. "I felt like I was a balloon someone blew up and let go." When the writer asked how many people had flown a rocket belt successfully, without tethers, someone told him that the number at that point was eleven. "More men have walked on the moon."[7] This statement was often repeated but the math was wrong. Twelve men walked on the moon; at that point there had been at least fourteen rocket belt pilots: Hal Graham, Peter Kedzierski, Bill Suitor, Bob Courter, Gordon Yaeger, John Spencer, Doug Meiklejohn, Donald Hettrick, Dennis Commerford, Lee Person, Tyler Nelson, Kinnie Gibson, Eric Scott, and Dan Schlund.[8] Admittedly, it was a small group, but the roster was growing.

Larry Smith, who would write about the event for PopSci.com and *Slate*, studied the attendees of the event and was fascinated. The attraction was obvious: the rocket belt promised flight to the average person. But he looked at the rocket belts and saw how unwieldy and heavy they were. And decades after the invention of the rocket belt, they still measured their time aloft in seconds. Speaking of the attendees and their awe for the people who had actually flown he said, "It's a really small club. It fulfills everybody's dream to fly. Everybody *wants* to fly but very very few people *can* fly. It is the smallest club in the world." To him it was a mixture of nostalgia, remembering a time when Bell was making space-age, futuristic inventions, with a touch of tech geekiness thrown in by those who thought the rocket belt's best days were still ahead.[9]

Mac Montandon wrote a book called *Jetpack Dreams*, chronicling his quest to find someone who would let him try on a rocket belt, and included a section where he described his attendance at the convention. Later, he said he was inspired by the other attendees—the nonfamous ones. He spoke to many of them at the convention and elsewhere and was struck by how many of them kept plugging away at the dream of flight, even if it seemed like it was never going to happen.

> There are just too many things that can go wrong. But the fact that these guys didn't stop trying—that they continued to tinker and dream big—this idea was what, in large part, drew me to working on the book in the first place. It's not often one sees such pure examples of true and great passion (yes, at times it looks a lot like passion's twisted cousin, obsession, but even that can be a wonderful thing)—and I wanted to get as close to that feeling as I could. This is the spirit of Wendell Moore. And it's alive and well—despite the inherent dangers of the pursuit—all over the world.[10]

Gijsberts and Clough decided to hold another Rocket Belt Convention in 2007. This one was two days long, held in conjunction with the Thunder of Niagara Air Show at the Niagara Falls Air Reserve Station on August 11 and 12. Many of the same rocket people attended the second show. A good mix of people was bound to show up: one website said it would consist of "rocketbelt test-pilots, builders, and enthusiasts from around the world." It may have been a bit of an overreach however, when they suggested that "current day rocketbelt builders will tell how to build your own rocketbelt."[11] Ky Michaelson was there and he brought a workshop full of rocket-powered gadgets for people to look at, including a rocket-powered chair. Suitor and Graham were there as well, representing the Old Guard.

Eric Scott brought the Go Fast Jet Pack to put on some flight demonstrations. To get to the show, Scott had to have his rocket belt shipped by a cargo plane. Scott had found it more difficult to fly commercially with his rocket belt and equipment as luggage. "At first, it was kind of cool. You'd hear, 'Mr. Scott, please come to the ticket counter,' and someone would come up and ask, 'Your luggage, sir . . . what, exactly, is it?'"[12] Scott flew the Go Fast Jet Pack three times at the conference, twice before big audiences and once for the press.[13]

Scott told a local reporter that he came to the conventions not just to fly a rocket belt but also to meet the pioneers in the field and pay his respects. "I want to meet the greats that went before me, the first guys to get in their rocketbelts and head up. I just want to say, 'You guys rock,' and maybe show them my own pack."[14]

Hal Graham reflected on his time flying the rocket belts and his thoughts during the short flights. "Not the least of which are, at the end of these 15 seconds, am I going to be landing safely or still in the air? It had its dark parts, it had its brief glimpses of glory, but the overriding factor was always looking, always thinking, 'What can I do to make this a safe flight?'"[15]

Kathleen Lennon Clough enjoyed the conventions immensely. While she loved the rocket belts, she really enjoyed the connection they had to her father. Tom Lennon was the official Bell photographer during the time the belts flew and he had witnessed—and photographed—as many flights as anyone. The time he flew the nitrogen rig had placed him into an elite group and people who came to the convention realized how important he had been to preserving the history. Better, people had filmed him while he flew. Clough said, "I've seen pictures of my father flying for the first time just this year. I've had eight reels of film he shot given to me. It's a part of him I never expected to see again, and it's great to have it."[16]

Although everyone appeared to be getting along swimmingly in the rocket belt community, things were not as placid as they appeared at the conventions. That same year, the matter of Kinnie Gibson's trademarks—"Rocketbelt" and "Rocketman"—wound up in court. Eric Scott, who flew for Go Fast at the rocket belt conferences, had originally worked for Kinnie Gibson at Powerhouse Productions. Gibson sued Widgery and Go Fast, alleging that Scott had stolen trade secrets from Powerhouse and that Widgery had improperly used the name "Rocketman" on his website. Interestingly, the federal lawsuit did not name Eric Scott as a defendant.[17] After a year or so of wrangling, the district court dismissed the case. On the matter of the trademark, the court said the trademarks "have been widely used by others for an extensive amount of time. The Plaintiffs' marks, therefore, are not legally protectable."[18] Likewise, the court dismissed the remaining claims regarding trade secrets and unfair business practices. Powerhouse appealed the court ruling but the appeal was denied.[19]

Also in 2007, Widgery announced that Jet Pack International was on the verge of a breakthrough. He was going to unveil the T-73, a device that could fly for ten minutes. It would not be powered by hydrogen peroxide but by jet fuel. Widgery didn't say too much about the engine at first, but it sounded like it would have a jet engine. A writer quoted Widgery saying the new T-73 would cost $200,000. For that price, you would also get flight lessons. According to the article, Jet Pack International was taking pre-orders for the devices, which would be available in 2008.[20]

In a later article Widgery said that the new flying device was indeed going to be a jet pack; it would be powered by a small turbine. The article also gave some different figures for the expected performance of the new product. It would fly for nineteen minutes and be able to travel as far as twenty-seven miles and soar to 250 feet. This article gave a $250,000 price and mentioned that flight lessons were included, and said the device would be available in December 2007.[21]

Widgery also considered manufacturing and selling the hydrogen peroxide rocket belts to the public but thought better of it. "It's really expensive for someone to learn how to fly. The fuel is expensive, and they're dangerous to the pilot until he learns how to fly." The rocket belt would have cost more than $100,000 and then the cost of the fuel during training would have been quite expensive. "And no one is going to spend a couple hundred thousand dollars just to get hurt."[22]

Widgery's company wasn't the only one talking about jet belts. In January 2008, Thunderbolt—Nino Amarena's company—issued a press release describing its progress in creating a rocket belt "for the masses." While the ThunderPack rocket belts were available, the company had now set its sights on jet power also. The ThunderJet would have "exponentially greater capability" than even the best rocket belts available. The new design might include two jet engines and allow for a thirty-five-minute flight time. Even so, this giant leap in technology would not cost much more than the rocket belts Thunderbolt was selling. Amarena stated that the ThunderJets would sell in the $100,000 range.[23]

By 2012, the ThunderJets had not been unveiled and the price of the rocket belts had crept up a bit. Even this field was subject to inflation. In 2012, anyone who had the money could buy one of Thunderbolt's rocket belts for $250,000. Amarena was careful to point out that a prospective buyer must make a choice: take flying lessons from Thunderbolt before taking delivery of the belt or sign a waiver, releasing Thunderbolt from any liability or harm that befell the untrained rocket belt pilot.[24] For the hopeful rocket belt pilot, Bill Suitor wrote a book that explains how rocket belts work and even gives a step-by-step flight training outline. The book is full of detailed photos of rocket belts and an amazing array of photos of Suitor in flight, many of which had never been published before. He told his story, the story of the rocket belt, and then how to maintain and operate one. To people like Peter Gijsberts, the book was a long time coming. Even so, Suitor cautioned his readers: "NEVER attempt to build your own rockets. This book is in no way intended to be a guideline to attempt such a thing. Only highly trained aerospace experts, working in controlled conditions can do this safely; finishing college will help prepare you for this type of job."[25] Suitor's warning fell on mostly deaf ears; civilians continued building rocket belts.

Bill Houghtaling of Florida began building his in 2008. He ran a pool maintenance business and had flown radio-controlled airplanes as a hobby. Houghtaling set out to build two rocket belts after talking about the belts with a fellow RC hobbyist. His friend put up some money, and Houghtaling did the work. He had been fascinated by the rocket belt since he saw one fly in *Thunderball*. When he was much younger, he had sent some money to an address in the back of *Popular Science* advertising "Build your own rocket belt!" In return, he got photocopies of the patent drawings for the belt—which were interesting but were not exactly a do-it-yourself outline for building a belt.

Houghtaling connected with many of the other rocket belt community members and began looking for expertise. Jeremy McGrane, who built the Go Fast Jet Pack, agreed to help. The two never met face to face; Houghtaling would take his own measurements and send them to McGrane, and McGrane extrapolated the data and made the various parts to correspond to Houghtaling's size. He bought his tanks from Juan Lozano in Mexico. He built his belt a piece at a time on nights and weekends and, a couple of years later, his belt was completed. It weighed around sixty-five pounds; he had used carbon fiber whenever possible. Houghtaling then ran into the same problem everyone else did: the hydrogen peroxide fuel shortage. He could find it, but it was expensive. Houghtaling shopped around for the machines to make the fuel but they were too expensive.

By 2012, Houghtaling had made tethered hops with his rocket belt. His only limits were a matter of time and the expense of the fuel. He has plans to master free flight and to perform rocket belt demonstrations. Chances are good the community will see him overhead one of these days.[26]

In 2009, *Popular Mechanics* published an article comparing three competing rocket belts made by Lozano, Amarena, and Widgery. The belts weighed between 139 pounds on the low end and 182 pounds at the heaviest. All were a bit unwieldy and new owners could expect skinned knees or worse while learning to land. Thunderbolt, Amarena's company, claimed a seventy-five second flight time for its belt, but the figure was an exaggeration. The others claimed to fly just a little better than the old belts, clocking in closer to thirty or thirty-three seconds.[27] *Popular Mechanics*—keeping in line with previous articles about individual lift devices—did not test or observe any of the belts to confirm the figures; they apparently just took the word of the manufacturers.

And while the belts were for sale, not many of them had sold yet. At least, no one would give the name of an actual buyer and no one ever showed up in public flying one of these belts, other than the makers themselves. Lozano turned down a potential buyer who wanted him to sign an agreement that looked like a lawsuit in the making. Amarena conceded that rocket belts weren't all that practical. Each thirty-second ride could cost hundreds of dollars in fuel depending on your source and, of course, the fuel was highly volatile and not something you could just leave lying around. Amarena said, "The rocket fuel is very finicky. You don't want it sitting around under your bed." It would be a slightly different world if jet belts ever became available. Jet belt fuel—essentially kerosene—was inexpensive and no more difficult to locate or handle than gasoline.[28] Hydrogen peroxide cost anywhere from $60 to $250 a gallon depending on whether you were making it yourself or buying it.[29]

And there was an even bigger problem buying it commercially after an industrial accident in Sweden. On July 22, 2010, there was an explosion at the Peroxide Propulsion manufacturing facility in Gunnilse run by Erik Bengtsson. Around 10:30 that morning, calls flooded the local fire and police departments as people reported hearing explosions and seeing a massive smoke plume above the rocket fuel factory. Bengtsson and two others were injured severely and had to be hospitalized. Two people nearby were also hospitalized with smoke inhalation. Fire crews later reported the source of the fire as aircraft fuel, forty cases of it, that caught on fire.[30]

Bengtsson said the incident had started out as a routine procedure. It ended with the founder of the company in the hospital and the factory, and its contents, burned to the ground. Although he was temporarily blinded by the accident, Bengtsson made a full recovery but decided to discontinue the business.[31] It was an indication of how difficult the fuel was to obtain when customers in America had been willing to pay a small fortune to have the fuel shipped over from Sweden. After this event however, it became even scarcer. Before the explosion destroyed his factory, Bengtsson had furnished his fuel to six or seven different well-known rocket belt operators on different continents. His son had come aboard the business as an equal partner. The fuel from Peroxide Propulsion had been used by rocket car drag racers, universities, and space agencies, and a host of people building rockets—the kind that shoot straight up into the sky with the goal of reaching outer space.[32]

By 2012, hydrogen peroxide was scarce. Several people in the rocket belt community were manufacturing their own fuel as a matter of necessity. Few suppliers of high-grade hydrogen peroxide were willing to sell to rocket belt operators.

Meanwhile, many of the rocket belt companies continued working on the next generation device: the jet belt. *Popular Mechanics* asked Troy Widgery of Jet Pack International about the proposed jet belt, the T-73, which had not appeared on schedule. "We were trying to have it by the beginning of 2008, but it could be the middle now."[33] The middle of 2008 came and went without a new jet belt. The *Popular Mechanics* writer pointed out:

> This is standard practice in the rocket-belt industry—announce one product, but divert attention to the better one down the line. And between Jet Pack's missed T73 deadline and Thunderbolt and TAM's vague promises, are we getting any closer to a viable product?[34]

The *Popular Mechanics* writer wondered if the rocket belt builders might have lost interest in the Bell-style rocket belt. "To their credit, the heads of all three companies are quick to acknowledge the relative uselessness of rocket belts. These are simply proof-of-concept designs on the way to the real prize—something you can fly for minutes, not seconds."[35] The writer suggested the companies were all offering to sell archaic technology while developing the advanced technology. But if the new technology was just around the corner, why would someone pony up a hundred thousand dollars or so to buy the old stuff?

Some people wondered if even the jet belts would sell. In 2012, John Spencer, one of Bell's test pilots, wondered about the market for a jet belt. "It's just too expensive." Beyond the curiosity element, he couldn't see why someone would be willing to pay several hundred thousand dollars for a jet belt. "If you have unlimited resources and you've got the bug to be able to do it, I guess you can go out and spend a million dollars. Hire a crew of three people to help keep it running for you and have a good time. But how many of those can you sell?"[36]

Thunderbolt Aerosystems' Nino Amarena was running into the same hurdles everyone else had when they tried to bring the jet belt to market.

In 2012, Amarena said he had located a supplier who could provide a jet engine capable of powering a jet belt and meeting or exceeding the performance of the Williams-powered Bell jet belt. The name of the supplier and the specifications for the jet belt were still being kept under wraps, but Amarena pointed out at least one big question mark: no one knew how the FAA would weigh in on this issue. How would they classify the device? What kind of requirements might there be with rating the engines for commercial sale to the public? Amarena was optimistic but pragmatic. A lot of work remained, but he insisted Thunderbolt Aerosystems would succeed.[37]

Meanwhile, the Go Fast Jet Pack—a hydrogen peroxide powered rocket belt—has been globe-trotting, much the way the old Bell Rocket Belts did in the 1960s. By 2012, the Jet Pack International team had been hired to make over six hundred flights in places as diverse as Switzerland, Brazil, Canada, Ecuador, Bolivia, and the United Kingdom. In Australia the Go Fast Jet Pack flew over the F1 Grand Prix. Flying the belt was a twenty-three-year-old named Nick MacComber, and he kept up a packed schedule. The Go Fast Jet Pack was routinely booked to fly ten to fifteen times per month.[38]

17

DUCTED FAN LIFT DEVICES

While inventors kept pushing to get a few more seconds of flight from the rocket belt, others were still looking for alternative ways to achieve individual flight. If hydrogen peroxide power was a dead end and the jet engine was too expensive, what avenues were left to pursue? Some inventors began experimenting with individual lift devices driven by propellers, in a sort of cross between the original Hiller flying platforms and the rocket and jet belts. One of the first to try this was a man named Michael Moshier, who founded a company called Millennium Jet in 1996. Along with an engineer named Robert Bulaga, the company secured funding from DARPA to create a state-of-the-art individual lift device. Millennium Jet wanted to power the unit with a gasoline engine that would run two ducted fans above the pilot just outside of and behind his shoulders. The 325-pound weight of the engine and fans would be too much for a person to support so instead of a backpack-like device, the unit would become a sort of exoskeleton the pilot would strap himself into. Still, the technology lifting it—gasoline engine and fans—was simple and largely off-the-shelf.

Studying the problems of individual flight, the engineers knew there was a problem with any heavy vehicle that spent much of its time hovering. The smaller of the two Hiller platforms had been stable but the larger one had been unruly and harder to maneuver kinesthetically. Millennium built a version of what they called the SoloTrek and tried controlling it kinesthetically. It quickly found that the device was impossible to control with the pilot's body movements. "We tried it initially and those were very interesting first attempts at flight," Bulaga said later, laughing. He found that the device was unstable and that a pilot—like so many inventors before him, Bulaga was piloting the device he designed—could not make corrections to the controls quickly enough to stabilize it. Millennium even hired an astronaut, one of the best pilots it could find, to see if the issue was pilot skill. It wasn't.

Bulaga designed a stabilization system into the unit so that it would correct itself in flight. This would become a "smart" individual lift device. He found a helpful physics department at Stanford and began making sensors to measure minute changes in direction and attitude and an onboard computer that quickly adjusted the SoloTrek's control surfaces to correct the flight. Such sensors are commonplace today, found in many smartphones and computer tablets, but they were not commonplace when Bulaga started working.

In essence, the system looked for changes in position that were not caused by the pilot—the computer monitored the controls to see if the pilot had asked for something to happen—and eliminated unwanted movements with counteractions. The SoloTrek was controlled by two handgrips that resembled the controls of the rocket and jet belts to a degree, but they controlled much more. The right hand control handled fore and aft motion as well as left and right banking. Twisting the control changed the yaw. The left hand was the throttle. This was a change from the previous devices, which had mimicked motorcycles with their right-handed throttles. The complex controls meant that a pilot would need training before simply strapping into a SoloTrek and lifting off. Bulaga suggested a trained pilot could learn to fly the SoloTrek within an hour. A nonpilot might need ten or twenty hours of flight training before being able to safely control the vehicle.

Millennium built a prototype, dubbed the SoloTrek XFV—for Exoskeletal Flying Vehicle—and flew it on a tether. It was powered by a one hundred horsepower two-cycle Hirth engine. The developers experimented with various engines and built a second XVF with a Wankel-type rotary engine manufactured in England by a company called U.A.V. Engines, Ltd. The goal, as always, was to find the lightest, most powerful engine possible. Bulaga flew the second SoloTrek on a safety tether in

December 2002, and it flew a little better than the first. The computer stabilization system operated, though Bulaga hoped to fine-tune it a bit. But the deadline to fly the SoloTrek for DARPA was approaching.

A few days later Bulaga tried another tethered flight and one of the safety tethers got too close to the top of one of the ducted fans. Faster than anyone could react, the duct ingested the safety tether. Instantly, the blade in the fan shattered as it diced the tether. Bulaga was only a few feet up but he still hit the ground uncomfortably as much of the machine came apart. Just as Bell had begun to worry about Wendell Moore hurting himself, some of Millennium Jet's investors began to question the wisdom of having the project's chief engineer conducting test flights.

The crash of the SoloTrek did more than damage one flying device. Millennium had promised to demonstrate the device in free flight, and the DARPA deadline for the flight passed before the device could be repaired. Further DARPA funding was now out of the equation. Millennium Jet had previously changed its name to Trek Aerospace—people had been confused by the name and assumed it built actual jets—and after the DARPA funding was taken off the table, Moshier left and the company reorganized.[1] Millennium received a small bit of good news. It had filed for a patent on the Single Passenger Aircraft in 1998; the patent was granted on December 3, 2002.[2]

Shortly before Moshier left the company, he spoke to a reporter about the progress the company had made with the SoloTrek. In the long tradition of inventors in this field of aviation, the hyperbole began almost before the device took flight. An article in the *Guardian* touted the success of the SoloTrek's flight. Downplaying the two-foot altitude that had been reached, the article noted that the device "should" be able to fly at seventy-five miles per hour for two hours and that "there is even a retinal scan device in the control helmet which confirms the pilot's identity."[3] It is unclear why such a thing would be needed, even if it did exist. Unfortunately for Millennium Jet, a NASA spokesman told a reporter, "It may turn out that the SoloTrek is not feasible, yet it will take us a step closer to a more feasible system."[4]

When the company reorganized as Trek Aerospace, Bulaga was made president. The SoloTrek was donated to the Hiller Aviation Museum, and Trek Aerospace began working on a line of what it called Springtail Exoskeleton Flying Vehicles. These used the same 118 horsepower rotary engine as the second SoloTrek did, but they had larger fans and ducts. The first vehicle of this type, dubbed the EFV-4A, flew on November 23, 2003, and was considered a success. The device was flown for a little more than an hour that day, testing the unit's stability. Trek Aeropsace was not shooting for any records but wanted to study the new device's improved flight characteristics. The unit had no trouble staying stable while aloft. It hovered for more than four minutes without the pilot touching down.[5]

Building on that success, Bulaga designed the improved EFV-4B. The 4B was said to be capable of traveling 184 miles in one flight. It could cruise at ninety-four miles per hour and stay aloft for more than two hours, carrying ten gallons of fuel at launch—at least, on paper. It weighed 375 pounds empty; that is, without fuel or a pilot. The numbers were calculated by computer models and had not been achieved in real life yet; again, the traditional practice in the industry.[6] It also had a theoretical flight capability of twelve thousand feet in altitude and carried a ballistic parachute.

As of this writing, the EFV-4B was not flying. Trek Aerospace was still developing the technology but focusing more on unmanned vehicles. It last flew a manned device in 2005. Trek Aerospace had the capability of building the manned devices, but it was unclear what their market would be. Bulaga later said the developers were "shooting to get the price under $100,000" when the company was considering manufacturing and selling the vehicles. Trek had pitched the idea to law enforcement, and state police in one state liked the idea. A trooper covering a multi-county area could get across a rural swath of countryside much faster if he could just travel as a crow flies.

Trek had spent $5 million from DARPA and $1 million in private funding over the entire program. They built two Springtails. One was sold to a businessman in San Diego who planned to display it at his place of business. The other one was dismantled for parts and experiments. Now, Trek Aerospace focuses on consulting with companies dabbling in ducted fan technology.

Trek was not the only company working with ducted-fan personal lift devices. In March 2005, a New Zealander named Glenn Martin filed a patent for a Propulsion Device in New Zealand and then, in October, in the United States. His invention was a dual ducted-fan lift device, very similar in appearance and operation to the Millennium Jet and SoloTrek aircrafts. In the preamble to the Martin Propulsion Device patent application, Martin wrote:

> Personal flight devices were developed in the 1960s and 1970s, but were essentially rocket based devices (jet belts) which gave extremely short flight times (typically about 26 seconds) and were difficult to control. Further, these devices were fuelled by rocket fuel which is intrinsically dangerous.[7]

Martin had confused rocket belts with jet belts and didn't appear to understand how they had worked. Controllability had not been the issue that made the rocket belt impractical; it had been the short flight time. And the jet belt didn't have that limitation; the drawback with the jet belt was simply the cost of the jet engine. Nevertheless, the US Patent Office granted the patent even though one would be hard-pressed to see differences between the Martin patent and the Moshier patent. The one major claim Martin appeared to be making about his device was that the blades in his fan spun in the same direction. He argued that any resulting loss of control would be made up for with the simpler design.[8]

In July 2008, Glenn Martin unveiled his device and announced he was calling it the Jetpack. It was a misnomer, since his device did not contain a jet engine—nor did it resemble any of the other devices people commonly called jet packs. It appeared he was simply trying to capitalize on the name recognition. His device was powered by a small engine driving two ducted fans.[9] It looked a lot like the Trek Aerospace Springtail. Martin called it the "world's first practical jetpack." At that time, he had only conducted manned flights at an altitude of three feet, but he told the press, "If you can fly it at 3 ft, you can fly it at 3,000."[10] The unit was designed to be able to fly more than thirty miles and Martin claimed it would hit speeds of sixty miles per hour. At least, those were the figures Martin gave to the press.

Martin unveiled his Jetpack at the Experimental Aircraft Association 2008 Oshkosh air show. Martin fired up the unit with a pilot and hovered it for less than a minute. During the entire flight two people held onto the device, suggesting it was not stable enough to hover for a short period of time on its own. Many of those present were disappointed because they were hoping for a jet belt display of old. Martin insisted he was simply being careful in his early development and that the device would eventually be proven viable. A year or two later, he put a test dummy in the Jetpack and, using remote controls, flew it up to five thousand feet above sea level. Once there, he shut the engine off and fired a ballistic parachute, much like the one that had been fitted on the WASP II. The Martin Jetpack parachuted safely to the ground, with almost no ill effect to the Jetpack or its dummy pilot. Some critics wondered if his device could fly out of ground effect with an actual person at the controls, more than a few feet off the ground. Some also wondered if the parachute test had been conducted with a dummy that weighed less than a typical pilot.[11]

Martin managed to gain quite a bit of publicity for his device, largely by doing what so many others in the field had done before him. He made extravagant claims about the practicality of his device and exaggerated to reporters. He told *Popular Science* that when his Jetpack hovered for forty-six seconds at Oshkosh, it "broke a long-standing record."[12] Martin appeared to be saying that his jet pack broke a record because it flew longer than the Bell Rocket Belt. Of course, he was comparing apples to oranges. *Popular Science* had reported decades earlier about the Williams-powered jet pack, which often flew longer than forty-six seconds. The record couldn't have been for a bladed device: the Hiller and de Lackner devices likewise flew longer than forty-six seconds. The de Lackner had once flown for forty-three *minutes*. It also couldn't have been for a dual ducted fan device like the Springtail by Trek Aerospace; they had flown theirs for longer than four minutes.[13] Never mind the fact that the forty-six-second flight had not been a free flight; it had been accomplished with two men holding onto the aircraft.[14] The writer of the article apparently did not ask Martin to explain any of this.

The Martin device weighed 250 pounds empty but could lift a pilot weighing up to 280 pounds. Some critics noticed that the Martin Jetpack was as large as, or larger than, many of the people who might pilot it. Martin had simply chosen to make tradeoffs in size and weight to gain reliability and economics. Instead of hydrogen peroxide fuel or an expensive miniature turbine engine, the Martin was powered by a two-liter, four-cylinder two-stroke engine that burned common gasoline. To shave weight from the device, Martin used carbon fiber and other modern lightweight materials wherever possible. The engine put out about two hundred horsepower, quite a bit more than the engine in the SoloTrek.[15]

At 250 pounds, the device was far too large for the pilot to actually support with his body. As a result, the Martin Jetpack rested on a built-in stand not unlike the exoskeleton of the SoloTrek; the pilot strapped into the unit and was then lifted by it. The pilot was overshadowed by the size of the device. The unit was also not kinesthetically controlled. Like the SoloTrek, it was a "fly-by-wire" vehicle where the pilot's control inputs were fed to a computer that then decided how the unit should be manipulated in response. The pilot worked two hand controls. The left hand controlled pitch and roll; the right hand controlled yaw and the throttle. Martin insisted that this was easier than kinesthetic control. "We recently had a group of absolute beginners all go solo within a few minutes of training and less than 5 minutes of flight time."[16] Critics remained skeptical. An untrained person flying something that is not kinesthetically controlled cannot be expected to react safely when confronted by an emergency in mid-air. With a telephone wire approaching, do you move the left handle to the right or left, or was it that you twisted the right handle?

Glenn Martin recognized there were some who thought his device was too large to be called a jet pack. "If someone says, 'I'm not going to buy a jetpack until it's the size of my high school backpack and has a turbine engine in it,' that's fine. But they're not going to be flying a jetpack in their lifetime."[17] Others appeared to simply be bothered by the fact that Martin had usurped the name "jet pack." Everyone knew that a "real" jet pack is small enough to wear on a pilot's back.

The Martin Jetpack also suffered from a bit of an identity crisis as a result of its design. In the advertising literature published by Martin, the jet pack was variously called "The World's Easiest to Fly Helicopter," "The World's Safest VTOL Aircraft," and "The easiest to fly Jetpack."[18]

One of the obstacles to selling a ducted-fan lifting device was the matter of how they would be classified in countries where they were sold. In the United States, the Federal Aviation Administration oversees aircraft and regulates what can be flown and by whom. Various news sources have reported that the Federal Aviation Administration classified the Martin Jetpack as an "experimental ultralight airplane," even though it would appear to more closely resemble a rotorcraft, like a helicopter.[19] With this classification it would not require a pilot's license to operate. However, FAA regulations would make the use of an experimental aircraft difficult.[20] Experts in the field also note a problem selling experimental aircraft to one market often regarded as ripe for such a device: fire and rescue. People familiar with aviation law note that it is not enough for fire and rescue people to want to buy the devices; organizations that bought them would also have to insure them, and very few insurance companies, if any, will insure an aircraft deemed experimental by the FAA, particularly when it is being used for a commercial or professional purpose.[21]

In 2008, Martin promised, "By next year, pilots will be soaring several hundred feet in the air, flying for 30 minutes at a time."[22] In 2010, *Time* magazine named the Martin Jetpack one of "The 50 Best Inventions of 2010." At that point, the devices still were not available for sale to the public. Martin continued development and kept promising. In February 2011, Martin said that it had been contracted to provide five hundred Martin Jetpacks per year to a "large overseas state enterprise for search and rescue."[23] Martin advertised that the jet pack would sell for $100,000 and to reserve one, you only needed to provide a deposit of $3,500.[24]

Martin was still in development with the vehicle in 2012. The company's website and advertising suggested Martin was on the verge of selling the jet packs to consumers but was simply ironing out the last details of the device.

18

HAL GRAHAM

Hal Graham was the Bell Rocket Belt pilot who demonstrated the device in front of President Kennedy and appeared in *Life* magazine, saluting the president with the belt on his back. After Graham left Bell he went back to engineering, working for different companies in the Buffalo area. Always a bit restless as an engineer, he went back to school and got a master's degree in accounting from the University of Illinois. He then moved to Syracuse, where he started his own CPA firm and taught at a nearby college. He dabbled in developing software in the 1980s and even did a stint as a justice of the peace.[1] People would hear about his past and ask: "You're the rocket man, aren't you?" Graham kept black-and-white photos of himself flying the rocket belt for those who asked.[2] When he retired he moved to Crab Orchard, about an hour west of Knoxville, Tennessee. He started a business as a charter pilot, shuttling people around in a small Piper aircraft. It wouldn't make him rich but it kept him busy and allowed him to fly, the thing he loved most. On the tail of his plane was a silhouette of a man flying with a rocket belt, a salute to his earlier days of flying when he didn't need an airplane to get into the air.

Graham attended the rocket belt conferences in New York in 2006 and 2007. He was invited by the organizers and he flew himself up in his own airplane. There, he let the attendees in on one of the best-kept secrets of the rocket belt program: he had crashed once and almost gotten killed. He told them about his last flight at Cape Canaveral and about how he had been knocked unconscious for almost half an hour. This little episode would appear to be a black mark against the 100 percent safe rating always given to the program, but by this time the program was nearing its fiftieth anniversary.

In 2009, the Federal Aviation Administration grounded Graham. He had suffered health problems, and after he recovered he had trouble with the exams administered periodically by the FAA. The FAA said he could turn in his license or they would begin proceedings to take it from him. He was seventy-five and didn't feel he had the strength or time for a fight.

On October 22, Graham drove to the FAA office in Nashville, where the decision to ground him had been issued. He walked into the building and, standing in front of a bank of elevators, pulled out a gun and shot himself in the head. Graham had no intention of harming anyone who worked there, but he was apparently making a statement. Back at his house, it was found immaculate, quite different than how the bachelor had usually left it. He had arranged his affairs and left Post-it notes to help the people he'd left behind. He left his will out where it could be found and wrote a long letter to his adult sons, who lived in Ohio and Connecticut. There was a stack of photos of him, saluting the president that day at Fort Bragg, wearing an army uniform with a rocket belt on his back. He signed and dated them.[3]

Hal Graham had been one of the biggest figures in the story of the rocket belt. Among the aficionados, he was often simply referred to as His Eminence.

19

JETLEV

When *Thunderball* came out in 1965, it was a worldwide hit. A thirteen-year-old at a screening in Hong Kong named Raymond Li saw the rocket belt and was hooked. "There were a lot of different gadgets in that movie but all I could remember was the jet pack. It was so cool," he said more than forty-five years later.[1] His family moved to Canada in 1969 and Li went on to school in Toronto to study science and business. Whenever he ran across articles about the rocket belt or jet pack he would study them, but he was disappointed to learn that the technology had not advanced appreciably beyond the twenty-one-second flight in *Thunderball.*

Around 2001 he rode on a personal watercraft and was impressed by how much thrust the device generated by pumping water. Could a similar system allow him to build a functional jet pack? As he thought about it, it began to make sense. He would build a jet pack driven by compressed water. The unit would be tethered to a small boat, which would feed compressed water to the jet pack on the back of the user. The engine and the fuel would remain in the little boat, allowing the jet pack to remain light. He mocked up a prototype and by 2005 had a unit he could fly over water for three minutes. Knowing that the rocket belts still measured their time aloft in seconds, Li felt he had a winner on his hands. Like the other designers of jet packs, Li couldn't help but test the unit himself. After the first successful hover test, he flew the next twenty-two test flights as well. He filed for a US patent and moved his company to Florida to continue working.[2] The preamble to his application read: "Personal flight has been an eternal dream and a recent reality. However, unlike birds, human beings have a low power-to-weight ratio."[3]

The controls on the device he called the Jetlev were similar to those of the rocket belt. The right hand worked the throttle and the left hand the yaw control, although his unit didn't use jetavators. He experimented with different linkage systems and finally settled on a fly-by-wire setup. The thrust nozzles on his device rotated, allowing for better control than the jetavators that only moved back and forth. In August 2008, the Jetlev was nearing perfection. He could take off and land in the water and was confident the unit was safe to be used by the general public. One of the biggest advantages of the Jetlev is that it operated over water, providing a safety net whenever it operated.

The downward water blast generated 420 pounds of thrust. The tether eliminated the hurdle faced by all of the Jetlev's predecessors: the unit did not have to carry its own fuel or a motor or engine. The support boat carried twenty-two gallons of fuel, enough for an operator to stay aloft for four hours. While the Jetlev was flying, the support boat trailed along behind, feeding fuel and water to the unit. Of course, that was the downside to the Jetlev. It remained tethered during its flight. The user of the Jetlev flew above the water and could fly in all directions, swooping and turning and zipping about, so long as the support boat was not too far off. Because of the proximity to water, Li had to design systems that could operate in and around a wet environment, which he did. Then, realizing that many of the users would want to hover over the ocean, he had to make sure the units were likewise impervious to salt water.

Li wondered about training: rocket belt pilots had all required dozens of tethered training flights before being able to safely attempt free flight. Li believed a person could be trained to fly the Jetlev and achieve self-controlled flight in one training session. For training, he built the unit with remote controls. When a new user first went aloft, the trainer stood by and controlled the units with the same kind of device used by a remote-control airplane pilot. The trainer got the rider/pilot comfortable with the unit and then gradually turned control of the Jetlev over to the person wearing it. What's the worst that could happen? The operator got wet. During his development of the Jetlev, Li invited Dan Schlund, the former rocket belt pilot for Powerhouse, to come down and try the device. Schlund was impressed. In many respects it seemed like a rocket belt. It just didn't have the frighteningly short twenty-one-second flight time.

The Jetlev also had another feature that made it easier to fly than a rocket belt. The water hose leading to the unit contained 150 pounds of water when the Jetlev was in flight, which acted to stabilize the device. A rocket belt pilot could be buffeted by wind gusts or turbulence. The same disturbances might hit a Jetlev operator but the tether would help counteract such sudden or unexpected forces.

In 2011, Jetlevs were put into use at various locations around the country. Li watched the operations closely to make sure the units were safe and to look for further refinements that could be made. In 2012, the Jetlev was offered for sale for $99,500.[4] While some might wonder about the practicality of the device with its limited range, the manufacturer was confident it could sell units to operators to sell rides on the Jetlev at resorts and vacation destinations, much like those who operate parasailing businesses in Florida or the Caribbean. People wanting to rent time on a Jetlev could do so in several locations in Florida as well as Hawaii, California, and Arizona's Lake Havasu. Li anticipated another market among yacht owners; a Jetlev took up about the same deck space as a personal watercraft. Raymond Li filed patents for his device in the United States and twenty-seven other countries. If nothing else, the Jetlev was probably the closest the average person would ever get to actually flying with an individual lift device.

20

JETMAN: YVES ROSSY

Yves Rossy was born in Switzerland in 1959 and is the man who may have brought the field of personal flight to its zenith. He piloted fighter jets in the Swiss air force and then became a commercial pilot. He loved to fly aircraft, but he wanted to fly himself. That is, to somehow remove the aircraft from the equation and make the experience closer to the personal flight of the individual lift devices. He met Arnold Neracher, the Swiss rocket belt engineer, and flew the belt that Neracher made, but he wanted to fly higher and longer. He spent more than ten years refining a carbon fiber wing three meters across, and he mounted two small jet engines on it. The engines were manufactured by a German company called Jet-Cat and were designed to power remote-control aircraft. The turbine engines were even smaller than the Williams engine in the jet belt: they were only five inches across and weighed a little over five pounds. The engines put out more than fifty pounds of thrust each and in 2012 they cost more than $5,000 each.[1] Rossy determined that the engines were powerful and dependable enough for him to use on his flights. It was obviously a long way for the technology to advance from the day when Thiokol had said they doubted anyone could build a practical small turbine engine.

Rossy's plan was to drop out of another aircraft and glide with his wing. He would augment the flight by firing the turbine engines. His wing would be kinesthetically controlled. He began working on his project in 1996 and decided his prototype was flightworthy in 2005. It worked, but he felt that the wing needed more power. He added two more engines and made the wing a little smaller. In November 2006, he flew over the town of Bex, Switzerland, and stayed aloft for five minutes and forty seconds. When his flight ended, he deployed a parachute and drifted to the ground with the wing still on his back.[2]

Rossy continued refining his wing and then set out to fly in some exotic places. By now, the press was calling him "Jetman." In 2008 he rode an airplane up to seventy-five hundred feet over the Alps with the wing on his back. He stepped out of the plane and fired the four turbines. Soon, he was zooming through the air at 190 miles per hour. "Steering with his body, Rossy dived, turned and soared again, flying what appeared to be effortless loops from one side of the Rhone valley to the other." Rossy appeared to have mastered the flying wing. At one point he flew horizontally and did a roll onto his back and then completed the roll, all under perfect control. "It's like a second skin," he said of the device. And it operated kinesthetically. "If I turn to the left, I fly left. If I nudge to the right, I go right." He ended his flight by parachuting down to the shore of Lake Geneva where he told reporters he planned to fly across the English Channel next and then the Grand Canyon. The refinements he had made to his four-engined wing made it weigh 110 pounds. By this point, Rossy said he had spent several hundred thousand dollars developing his wing, and he was not optimistic about it ever becoming a consumer device. [3]

In September 2008, Rossy flew across the English Channel. With National Geographic filming, Rossy stepped out of an airplane at eighty-two hundred feet and fired his turbines. His twenty-two-mile trip began above Calais and took him just thirteen minutes as he reached speeds of 125 miles an hour. When he got over the Dover cliffs, he did a couple of loops and deployed his parachute. When he landed, he spoke to reporters. "With that crossing I showed it is possible to fly a little bit like a bird."[4]

Rossy had spent fifteen years by then working on his wing and, much like the rocket belt builders elsewhere, had done much of the work in his garage. Rossy's wing was a bit more high-tech than the typical garage-built project, though. The four engines needed to keep running at the same time; there would be a major problem if one quit unexpectedly in midair. Rossy's wing was equipped with digital processing equipment that analyzed engine performance. If an engine died, the unit would shut off the corresponding engine on the other side of the wing within a half a millisecond. "Otherwise, his wing would go into a flat spin," one of his component manufacturers told the press.[5] National Geographic explained to its viewers that Rossy's wing was made from carbon fiber, glass fiber, and Kevlar. The Kevlar covered the turbines to contain any shrapnel if they were to explode in flight. Without this "bulletproof vest"—what his engine manufacturer called it—Rossy could be injured by flying engine parts in the event of a failure.[6]

In November 2009, Rossy had a rare failure while attempting a flight over the Strait of Gibraltar after being dropped from an airplane high over Morocco. The flight started well at sixty-five hundred feet over Tangier but shortly into the flight one of his engines stalled. Rossy simply cut the wing loose and he and the wing each parachuted into the sea. He got a ride to Spain in a helicopter, which took him to the hospital. He appeared to be unhurt, and the Spanish Coast Guard retrieved his wing for him.[7]

In 2011, Rossy decided to fly over the Grand Canyon. He contacted the FAA to find out what issues they might have with his flight. The FAA admitted it was unsure of how to classify Rossy's Jetman wing. It gave him a list of things he needed to do for safety, and Rossy began preparing. The FAA waived the minimum flight training requirements for Rossy after inspecting his suit and his flight plan.[8] But he was not prepared for the onslaught of the US media. Journalists and cameras appeared from all over to film the flight, which was sure to prove breathtaking. Rossy's earlier flights over Europe had been making headlines everywhere and the films were becoming Internet hits. Now he was tackling one of the most spectacular venues in the United States. The day of his flight, he called a press conference and announced its cancellation. "If I do [make] a mistake and half of United States television [is here], it's really bad for you, for me, for everybody. I don't want to take the risk to present something unprofessional." He said he was worried he might not have trained enough. He gave some interviews and said he would reschedule the flight. The media eventually wandered off.

The next day, without making any public announcement, Rossy climbed aboard a helicopter and headed up to a height he could jump from. Under a clear blue sky, he let go of the helicopter and fired his four jet engines. He soared along the canyon for eight minutes, sometimes just a couple hundred feet from the rim and at speeds approaching 190 miles per hour. The Swiss watch company Breitling was now his major sponsor and its name was emblazoned across his wing and on his flight suit. The event was recorded by several cameras, including one on Rossy, and he uploaded it later to the Internet. Rossy said he hadn't been trying to mislead the press. He had been waiting for ideal weather conditions and when they arrived, there wouldn't have been time to wait for the press to show up if an announcement had been made. Some news sites suggested Rossy had been spooked by the amount of media attention but it seemed just as likely that Rossy's sponsors wanted to film the event themselves and keep it exclusive.[9]

From this point, Rossy's flights became more spectacular and were always well documented. In May 2012, he headed for South America. Over Rio de Janeiro, Rossy stepped from a helicopter wearing his wing and spent more than eleven minutes flying by and around the landmarks of the city. He flew over Ipanema and past the Christ the Redeemer statue. At the end, he cut the engines and deployed his parachutes. Rossy had built the unit its own parachute and Rossy floated to the ground without the weight of the wing on his back. He landed on Copacabana beach. His sponsors covered the event with several cameras and soon uploaded video of his flight to the Internet.[10] A few nights later, he appeared on *The Late Show* with David Letterman. Rossy explained to Letterman that the wing had a four-to-one glide ratio and could be built for about $100,000. He told Letterman that the next goal was to get bigger engines and fly "vertical." As he said it he motioned with his hand, pointing straight up. Letterman commented: "You're the luckiest man in the world."[11]

By mid-2012, Rossy said he had developed his wing to the point where he could launch from the ground—if he wanted to. It wouldn't launch vertically: he'd have to use a runway and some sort of wheeled undercarriage that would allow him to accelerate to flight speed. Then, as he lifted off, he would let go of the undercarriage and fly as he does now. Rossy considered this way of launching too dangerous to try, however. It would also burn an inordinate amount of fuel just to take off. Then, in the moments after he had let go of his undercarriage, he would be in grave danger if anything went wrong. Much like a passenger jet's most dangerous time is during take-off, Rossy would be exposing himself to unnecessary danger to launch this way. For now, he is content knowing his flights are higher, longer, and faster than those made by any of the other personal flight devices that preceded his. He was also training another pilot—a man with expertise in parachuting—to fly the wing. He hopes to be able to fly two wings in formation in the near future.[12]

And even if Rossy took off from the ground, he still needed to land by parachute due to the high speed at which his wing flies. It would be impossible to come in and land on the ground at such speed. It is unlikely that average people will ever be flying like Jetman Rossy to avoid the traffic of the morning commute. Still, his widely publicized and well-documented flights have shown us that he can fly great distances and at great speed, using nothing but his body movements to control his flight. Once again, the dream of personal flight is tantalizingly close. And still, it remains just out of reach.

21

WHEN WILL WE HAVE JET PACKS?

So how close have inventors come to the notion of individual flight? What have we learned as man has experimented with the various devices since the Zimmerman Flying Shoes and the Thiokol jump belt? For a man to fly safely, he must find a power plant strong enough to lift him, itself, and the fuel it will run on. It is not enough to simply strap on a rocket and blast off; to survive the flight, the pilot must also be able to control his flight while being carried by this engine.

Rocket motors work because they are light enough for a pilot to wear on his own body and they allow for kinesthetic control. This last idea makes the devices simple; you do not need to be a pilot to fly a rocket belt safely. It also means that the device is lighter. Controls, regardless of how simple they are, are bound to weigh something. On the other hand, rocket motors burn fuel at a rate that makes it impossible for a pilot to carry more than a fraction of a minute's worth of fuel. Using more powerful fuel would be an approach to this problem but—as we saw with the isopropyl nitrate–powered Sud Ludion—even those improvements are only incremental. What if we could find a fuel two or three times more powerful than what we have? A rocket belt could fly for a couple of minutes then. We still would not have practical individual flight.

Turbine engines have a good horsepower-to-weight ratio and burn fuel much more slowly than the rocket belts. Flight times are longer, perhaps in the ten- to twenty-minute range, but the engines have been prohibitively expensive so far. Rumors swirl of companies that may offer jet belts for sale in the $100,000 to $200,000 range, but so far no one has been able to find a supplier of engines that could bring the cost down to that level.

Rossy's jet-powered wing appears to have found the perfect balance of power and weight. He can manage to handle it and it flies—so long as he can be dropped from an altitude to get going. He could launch it from level ground but it would not be a VTOL aircraft—the Jet-Cat engines he uses are not powerful enough to provide vertical liftoff.. He would need a runway and would have to use some sort of landing gear during take-off. While in the air, his device is a wonderful personal flight device, but it still needs a parachute to land.

There is always the problem of emergencies. A jet belt or wing could carry a ballistic parachute that would help in engine failures above seventy-five feet. Below that height there could be problems. Of course, this would mean more weight for the device and no guarantees of safety, considering how often a person would need to be flying under that seventy-five-foot ceiling.

The piston engine is less expensive and requires less maintenance than a turbine, but it is heavier. One powerful enough to lift a man is too large for a pilot to comfortably support. Instead of being a backpack-like device, it requires an exoskeleton-like device. While these have been made, the size and shape of the frame means that it outweighs the pilot and makes kinesthetic control impossible. The device flies, and while it might not require a pilot's license, it requires some training. It is also larger and heavier than the other flying belts and would be regulated by the FAA. It certainly would not be something the average person could just strap on and fly.

What about the flying platforms? The ducted fans of the 1950s showed promise. They were relatively simple and used piston engine technology. But they were not terribly safe. Keeping two engines running in synch is a tricky proposition and the failure potential of the counter-rotating propellers is also a concern. Any steps taken to make the system safer, such as replacing the belt drives with a chain drive or adding a third motor, add to the complexity and cost of the device, and make it heavier and harder to fly. Of course, if these were to fly more than a few feet off the ground, the FAA would impose aircraft rules on them. Would people buy something that was not much more than a Segway riding a foot or two off the ground? Such a vehicle would not fulfill the promise of being able to fly in the traditional sense.

And all of the devices come with the problem of "flight degradation." That is, what happens if the device loses power? The best-case scenario involves a ballistic parachute and the operator hoping to find a clear place to land. An engine failure closer to the ground would mean serious injury and possibly death. Of course, the falling individual lift device could harm someone else, depending on where it fell and who it hit on the way down. To date, no one has even suggested a solution to the stalled-jet-pack problem other than a ballistic parachute.

It seems that the notion of personal flight is just out of reach, as it has always been. It is like a mirage; as we get closer to it, it seems to move further away. It seems there are just two or three variables—when two are solved, the third becomes insurmountable. Solving the third issue causes one of the first two to go out of whack and so on.

And even as inventors wrestle with these problems, other hobbyists continue to build and fly the hydrogen peroxide rocket belts, technology that was largely perfected in 1961. A former vice president of Bell, Hugh Neeson, went to the two rocket belt conventions and was asked about all the attention still being given to the rocket belt. He found some of it hard to process. In the decades since the first belt flew, the rocket belt pilots were still working with thirty seconds or less of flight time. "The inherent laws of chemistry and physics of the question . . . of keeping a man in the air under the pure power of thrust, that tells you that you can't get a lot more duration out of this. The effort being invested in [the rocket belt], though, is a bit puzzling."[1]

Will we ever have affordable personal flight? It has been more than fifty years since Hal Graham took the first step and it seems humankind has not gotten much closer in the ensuing half-century. Douglas Adams gave advice that might be the only practical advice for the average person on how to get off the Earth's surface:

How to Leave the Planet:

Phone NASA. Their phone number is (731) 483-3111. Explain that it's very important that you get away as soon as possible.

If they do not cooperate, phone any friend you may have in the White House—(202) 456-1414—to have a word on your behalf with the guys at NASA.

If you don't have any friends at the White House, phone the Kremlin (ask the overseas operator for 0107-095-295-9051). They don't have any friends there either (at least, none to speak of), but they do seem to have a little influence, so you may as well try.

If that also fails, phone the Pope for guidance. His telephone number is 011-39-6-6982, and I gather his switchboard is infallible.

If all these attempts fail, flag down a passing flying saucer and explain that it's vitally important you get away before your phone bill arrives.

One final note: Regardless of anything else, there is one thing we can always count on. The media, led by *Popular Science*, will believe and reprint almost anything told to them by the builders of an individual lift device. They will not question claims of what a device will do. In fact, the day they do question an outlandish performance claim made by an inventor, it might be time to start looking for flying saucers.

EPILOGUE

WHERE ARE THEY TODAY?

The first Hiller Flying Platform is housed at the Hiller Aviation Museum in San Carlos, California. The Hiller 1031-A-1, the second Hiller Flying Platform built, is in the collection of the Smithsonian National Air and Space Museum. Currently, no one knows where the Flying Shoes are, or if they still exist. According to one writer, Hiller gave them back to Zimmerman in the summer of 1948.[1]

One de Lackner Aerocycle survives and is housed at the US Army Transportation Museum in Fort Eustis, Virginia.[2]

Bell created five rocket belts. Bell donated belt #2 to the Smithsonian National Air and Space Museum in 1973.[3] Another is at the US Army Transportation Museum in Fort Eustis. The University of Buffalo has one, and another is at the Niagara Aerospace Museum in Buffalo. Hopeful visitors to the museum are cautioned to check in advance to see if the museum is open; as of this writing, it isn't.[4] The fifth rocket belt was dismantled so that its parts could be used to create other rocket-powered devices like the Pogo. It was never reassembled.

There were two Sud Ludions built. One was destroyed. The other can be seen at the French Museum of Air and Space, near Le Bourget airport outside of Paris.[5]

The RB-2000 "Pretty Bird" is still missing. As late as 2007, Barker was still refusing to discuss the whereabouts of the device that, legally, belonged to Stanley.[6] Of course, Barker had never admitted if he still has it, or if he actually knows the belt's current whereabouts.

Only one jet belt was manufactured under the Bell contract and it is still owned by Williams International of Walled Lake, Michigan. It is not on public display at this time.

Three WASP IIs were built.[7] Williams International has one. Another is on display at the Museum of Flight in Seattle. One is suspended from the ceiling of the National Museum of the Air Force at the Wright-Patterson Air Force Base in Ohio. The Dr. Sam Williams Jet Age Gallery maintains videos of the WASP II in flight and biographical information on Williams and other important figures in jet technology. The location of this WASP II—suspended in the air—not only gives the appearance of it in flight, it allows visitors to get a good look at the underside of the vehicle and the vanes in the exhaust that controlled the yaw. Interestingly, this WASP II was originally painted olive drab for its military sponsors. It has since been repainted in the manner of its civilian X-Jet iteration.

Elsewhere in the same museum, visitors can see an F-107-WR-101 Williams engine, a descendant of the power plants found in the jet belt and the WASP flying platforms. Suspended from the ceiling as if flying overhead are two cruise missiles powered by Williams engines. The air-launched cruise missile is driven by a variant of the F-107 while the other missile is also powered by a Williams engine.

Two SoloTrek devices were built. The first is at the Hiller museum. The other was dismantled for its parts by its makers.

APPENDIX

ORIGIN HOAXES

Despite Wendell Moore's well-documented invention of the rocket belt, stories and folklore abound of inventors whose roles in the creation of jet pack technology were somehow overlooked or suppressed. The stories are not true but some of them have gained traction, particularly on the Internet. A brief search of the Web on jet pack history will turn up several websites that claim the Bell Rocket Belts were copied from a German secret weapon from World War II. One popular website ran a piece entitled "Real Bloody Flying Nazi Soldiers with Jet Packs."[1] According to this story, the Germans experimented by strapping pulse-jet engines onto soldiers and launching them over obstacles on the battlefield. The pulse-jet was the same engine that powered the V-1 Buzz bomb, and seems a highly unlikely thing to strap to anyone if you expected him to survive the experience.[2] Websites that tout this device do not offer any photographs of the Himmel Stürmer—which they point out meant sky stormer—and the story seems shrouded in mystery. The writers said that the Americans captured the technology along with some other secret weapons and brought some of the devices back to America. Some sites go so far as to say that Bell Aerospace obtained one of the devices, and Wendell Moore dissected it before coming up with his rocket belt. *The Rocketbelt Caper* devotes two pages to the mythical device with the caveat, "a lack of evidence suggests the Nazi rocket packs may have been pure propaganda."[3]

"Propaganda" implies that the Germans said they had one, something that is also not supported by any evidence.

The stories are fiction. The websites that describe the Nazi jet packs cite each other for support and when pressed, the authors of some of the sites admit they have no evidence that any such device ever existed.[4] One site sold dolls—"action figures"—made up to look like storm troopers wearing little jet packs. Photographs of the dolls appeared on another website as evidence that the Himmel Stürmer was real. One writer claimed that the lack of evidence on the device proved its existence; if it wasn't real, why had the US government hidden the evidence?[5]

People familiar with jet belt and pulse jet technology point out that the Himmel Stürmer could not have possibly been real. The engine in a V-1, known as an Argus As 014, was thirteen feet long. That would be a rather difficult thing to strap to one's back. Of course, the engine could be downsized, but then the resulting engine would have been much less powerful. Could this have been done?

Disproving negatives can be a difficult thing. In this instance, it makes more sense to try and track down the Himmel Stürmer story to its source. A writer using the name Christof Friedrich first wrote about the Nazi jet packs in a book called *German Secret Weapons and Wonder Weapons of World War Two* in 1976. He also wrote a book called *The Hitler We Loved and Why.* His real name was Ernst Zündel and he ran his own publishing company in Toronto. At first, many of his books, written under the pen name Christof Friedrich, were about UFOs and how the Germans had developed advanced technology during World War II that was kept secret from the general populace. Zündel explained how Hitler ran "secret Nazi polar explorations" in a book by that title in 1978, and wrote *Hitler at the South Pole* in 1979. Somehow, a few readers who didn't see the lack of credibility in Zündel's writing saw his German secret weapons book and believed it to be true. It wasn't. One writer summarized Zündel's theories:

> Some hollow Earth believers exhibit not just fascination with but open sympathy for Nazi Germany. The chief figure in the Nazi hollow-Earth movement is a Toronto man named Ernst Zündel, who writes under the name Christof Friedrich. Zündel operates a clear-

inghouse for Nazi materials and contends, as do other neo-Nazis, that the Holocaust never took place. In *UFOs – Nazi Secret Weapons?* (1976) he claimed that when World War II ended, Hitler and his Last Battalion boarded a submarine and escaped to Argentina; they then established a base for advanced saucer-shaped aircraft inside the hole at the South Pole. When the Allies learned what had happened, they dispatched Adm. Richard E. Byrd and a "scientific expedition" – in fact an army – to attack the Nazi base, but they were no match for the superior Nazi weapons.[6]

We do not know if the Nazi soldiers defending Antarctica after World War II flew Himmel Stürmers or not during their epic battle; Zündel didn't mention the devices in that story. Perhaps a bigger concern was that Zündel had an ulterior motive for spinning his crazy UFO stories. He was a rabidly anti-Semitic Holocaust denier who used the UFO theories as a platform from which to preach his theories on what "really" happened during World War II. Maybe his book *The Hitler We Loved and Why* should have tipped off his readers, but the other titles he published certainly should have, such as *Auschwitz, Dachau, Buchenwald: The Greatest Fraud in History*, and *Did Six Million Really Die?* Zündel was a bizarre character in that he openly spoke about his beliefs and granted interviews to almost any journalist or reporter who approached him. One such writer named Frank Miele interviewed Zündel in 1993 for an article in *Skeptic* magazine. The article focused on Zündel's revisionist statements regarding the Holocaust and Miele asked him about his UFO literature. Zündel told him that it was the people who believed in UFOs—not the Holocaust deniers—who were the "real lunatic fringe."

In a later phone conversation, Zündel told me that the UFO book was in fact a ploy. "I realized that North Americans were not interested in being educated. They want to be entertained. The book was for fun. With a picture of the Fuhrer on the cover and flying saucers coming out of Antarctica it was a chance to get on radio and TV talk shows. For about 15 minutes of an hour program I'd talk about that esoteric stuff. Then I would start talking about all

> those Jewish scientists in concentration camps, working on these secret weapons. And that was my chance to talk about what I wanted to talk about."[7]

The UFO book he was talking about was *UFOs—Nazi Secret Weapons?* Clearly, Zündel has no credibility regarding anything he wrote about these topics. He wrote the books because they sold and he used them as an entrée to talk shows because the topics were so popular. He had no evidence of UFOs, South Pole Nazi submarine bases, or of the Himmel Stürmer. He made them all up. The Himmel Stürmer was a figment of his imagination.

The Himmel Stürmer hoax was not the only false story about the origin of the rocket belt. Sometime before 2008, the Romanian National Geographic channel broadcast an interview with a Romanian inventor named Justin Capra.[8] Capra, who was born in 1923, explained how he had spent his life in Romania, building countless inventions.[9] He had created lightweight cars, voice-command remote controls, and experimental aircraft. And, he said he invented the jet pack in 1956. The show did not ask him to elaborate and apparently did not question any of his claims. Capra explained how he had conceived and built his "rucksack" flying device while in the military; he had been wondering how a soldier could evade sentries. He provided the show with black-and-white photographs of himself wearing a rocket belt–like device. To illustrate to viewers how Capra's jet pack flew, National Geographic inserted footage of a Bell Rocket Belt test flight.

In 2008, a celebration was held in Romania to honor the "fiftieth anniversary of the first flight made by individual flight device—Justin Capra." Film of the event can be found on the Internet, showing Capra walking through a museum displaying several of his inventions, two of which could be described as individual lift devices. One was a backpack with two orange rockets mounted on it. Another looked more like a rocket belt in that it was configured with exhausts that appeared to flow out of tanks the pilot wore on his back.[10]

After the fanfare of the fiftieth anniversary, Calin Dinulescu, a Romanian filmmaker, tracked down Capra to interview him further. Dinulescu spoke with Capra for several hours on camera and Capra elaborated on his claims regarding his individual lift device. The interview was then edited to less than four minutes and posted on the Internet. The introduction to the piece did not qualify any of Capra's claims and simply introduced him as having "designed seventy prototypes of fuel-efficient vehicles, seven unconventional flying machines and fifteen alternative engines. The most notable of which was the jetpack."[11] As with his jet pack, there was no evidence shown to prove that any of his unconventional flying machines ever actually got off the ground. Capra eventually gave numerous interviews to newspapers and magazines in Romania and many articles have appeared there about the Romanian inventor of the jet pack.

In the Dinulescu video, Capra claimed he did not patent any of his inventions. "I didn't even understand the meaning of the word 'patent' back then." He stated, "The first major device I designed was the personal jet pack, proposed at the Academy in 1956 and completed in 1958. I was thinking of a way to escape the barracks without being noticed." Fifty years after the events he was recounting, Capra claimed the Romanian Academy was not interested in his invention. They told him, "Comrades, we need tractors, not people to fly."[12] He then offered it to the US Embassy but they said they weren't interested either. When the Romanian authorities found out about his visit to the embassy, they threw him in jail. After he got out of jail, he spent two years working on the jet pack. After he had a working model—which he flew himself—he applied for and received a patent. But, since Romania was a communist country, all patents belonged to the people and Capra did not receive any personal credit or acclaim for his work.[13] The document Capra provided to the filmmaker was a single blurry page of a patent. It contained no drawings and the date was illegible. Romanian *National Geographic* published an article in 2011 stating that Capra had been awarded a patent on July 8, 1958, but they did not make any claims about whether he ever built or flew the device he patented.[14]

Capra provided several of his interviewers with old photographs showing one of the two devices seen at the fiftieth anniversary celebration. The device was not flying in any of the photographs, and it was unclear how he could prove when he built the machine or even when the photos were taken. In the Dinulescu video he claimed, "The first American design appeared on February 22, 1962, four years after ours." He was wrong on that: Hal Graham had free-flown the rocket belt in 1961. It may be that he was referring to the patent date, which was granted to Moore on February 13, 1962. Even so, Moore had applied for the patent June 10, 1960. Interestingly, he claimed the American jet pack was patented by three people: Wendell Moore, Cecil Martin, and Robert "Cuning." The Bell Rocket Belt patent bore the name of only one inventor: Wendell Moore.

What caused Capra to blame three Americans? A search of jet-pack-type patents in the United States turns up a patent for a Turbo Fan Lift Device invented by Cecil Martin and Robert Cummings. They did not work for Bell; they worked for a company called Thompson Ramo Wooldridge (TRW) in Ohio. And their device looked nothing like Capra's jet pack or Moore's rocket belt. They filed for their patent in 1958, the same year Capra said he applied for his in Romania.[15] Capra apparently pointed to them as having stolen his idea because their patent was *granted* around the same time as Wendell Moore's. However, the TRW patent had been *applied for* in 1958.

Dinulescu also inserted footage of a Bell test flight into his Capra interview—Bob Courter flying the Bell Rocket Belt through a slalom course—to illustrate how Capra's jet pack flew. "Anyway, it happened that we were the first ones to develop the design. The main difference between mine and theirs was the color." It is unclear if Capra ever saw a Bell Rocket Belt; if he had, he would know that there were major differences between the rocket belt and the device he touted as his. Among other things, the Bell belt had three cylinders whereas the Capra jet pack had six spheres to hold its fuel. They also appeared to be the same color more or less: the Capra jet pack was gray and silver, the Bell Rocket Belt was white, gray, and silver. In some interviews, he said the rocket belt was black.[16]

Capra told many of his interviewers that he also flew his jet pack without mentioning how he learned to fly it. Did he use tethers? How many test flights did he make before free flight? "When I lifted-off, I got scared. I didn't expect it to happen so suddenly, and my movements were made purely by instinct." How long did he stay aloft? How many times did he fly it? Why were there no photographs of him in the air? Why were there no contemporary newspaper articles about his invention? Capra never mentioned any of these issues. Even so, Capra's claim of inventing the jet pack was widely believed in Romania, though nowhere else. Many of the articles about him went so far as to say that "the Americans" finally recognized Capra as the true inventor of the jet pack in 2002.

How much truth was there to Capra's claims? First, it appeared that Capra designed two different jet packs, and the one he touted as his groundbreaking jet pack was the second one, built in the mid-1960s. Capra gave another interview in 2008 where an interviewer asked for some specifics about what he had done and when. He described his first invention as "two rockets arranged symmetrically on both sides of the shoulders." This dual-rocket design was also described in a 1995 publication that described Capra's jet pack as a "portable plane for individual flight, carrying two minirockets as a propulsion system. His invention of 1958 represented the grounds of the auto-propulsion systems used by spacemen in order to move outside the spacecraft."[17]

In the 2008 interview, Capra admitted his 1958 design did not work. "I only managed to make a small leap." His first jet pack was fueled by alcohol. He told the interviewer that "truer flights" were not made until 1968 after he reconfigured a jet pack to use hydrogen peroxide—like the Bell Rocket Belt.[18] He also admitted to seeing the Bell Rocket Belt after he had made his first jet pack but before he built his second.

Capra's second jet pack was displayed in the Romanian National Technical Museum in Bucharest. The sign on the device stated that it was invented in 1958 and that Capra had a co-inventor, Ion Munteanu. However, information tags on the machine indicated that the tanks on the unit were certified for use in September 1967, which would support the notion that the device was put together shortly before his tests in 1968 and that this was the second jet pack he built, not his first.[19]

There is no question that Capra built two jet-pack-like devices; one of them even resembled a rocket belt in many ways. It is unclear if either of them would work. Among other things, the exhaust tubes on the second jet pack appear to be too short and too far above the pilot's center of gravity to provide stable flight. The data plates on the second device indicate that the tanks have a three-liter capacity. That would mean the entire unit could hold less than five gallons of fuel. But it would still need the nitrogen—on the Bell Rocket Belt one of three tanks holds nitrogen—so four of Capra's six fuel tanks could hold 3.17 gallons. The Bell belt used six gallons to fly for twenty-one seconds. Assuming Capra's belt was capable of flight, it appears his device carried enough fuel to run for about eleven seconds.

Still, it is accepted as fact by many in Romania that at least one of Capra's jet packs was built and flown before Bell's.[20] Further, many of the biographies of Capra state that his jet pack was stolen by the Americans.[21] One example is the website for the Association of Romanian Engineers in Canada. On their website, under "Famous Romanians," they tell the story of Capra and his jet pack. It was one of his inventions, they said, worth "billions of dollars" that was "simply stolen by Americans."[22] They do not explain what other inventions of his were stolen by Americans.

Searches of the Romanian patent office's database revealed only two patents in Capra's name, neither of them describing a jet pack. If Capra built a jet pack of any sort in 1958, it was most likely one with two rockets attached to a backpack. This, in itself, would not bc groundbreaking. There were patents in the United States for rocket systems attached to humans as early as 1945. And, a patent is not proof that a prototype was built or that a design was viable. Capra apparently has no independent evidence that he built anything prior to 1967, which would have been more than five years after the Bell belts became widely known around the world. A better explanation is that Capra came up with an idea for a man rocket and filed for a patent on it in Romania. He may have been the first to do that *in Romania*. After he saw the Bell belts fly, he set out to copy the Bell belt. He built the device, his second jet pack, which was now in the museum, but he probably never flew it. One would think he would have documented such an important feat, if he had ever done it. Now, fifty years after the Bell belts first flew, he told his countrymen that he beat Bell by two years. He pointed to the date on the patent for his first, dual-rocket "jet pack" and then to his second jet pack, which was built a decade later, without any evidence to prove that they existed at the same time.

ACKNOWLEDGMENTS

One final note is necessary. Anyone researching any of the topics in this book will come across the name Peter Gijsberts sooner or later. Peter runs the website www.rocketbelt.nl where he has posted a vast collection of material relating to the field of rocket belts. As can be deduced from the Web address, Peter lives in the Netherlands and simply has a fascination with the rocket belts. While researching them himself, he made contact with many of the players in the story and even organized—with the help of Kathleen Lennon Clough, whom he met along the way—two rocket belt conventions in 2006 and 2007 in Buffalo. There, he met face-to-face with Hal Graham, Bill Suitor, and Peter Kedzierski. I met Peter by contacting him via e-mail and he offered me endless help. He answered questions, provided me with materials, pointed me in the right direction, and got me in touch with many of the people necessary for me to finish this book. This book would not have been possible without the help of Peter Gijsberts.

Otherwise, I need to thank: Robert "Bob" F. Courter, Bill Suitor, Evelyn and Weston Ihrke, Jay Spenser, Mark Voss, John W. Spencer, Troy Widgery, Ky Michaelson, Kathleen Lennon Clough, Juan Manuel Lozano, Carmelo "Nino" Amarena, Ray LeGrande, Scott Watson, Mark Voss, Jill Warner Dailey, Arnold Neracher, Steve Harris, Yuval Taylor, Diane Porter at the University of Buffalo, Raphie Aronowitz, Robert Bulaga, Benjamin Gilliand, Don Flanigan at the Hiller Museum, James Bowker at Martin Jetpack, Robert G. Loewy, John Armstrong at Wright State University Library, Doug Malewicki, Markku Jaakkola, Gregory Mone, Wes Gibson, Kinnie Gibson, Scott Decius, Brian Dailey, Colleen Keeler, Scott Reizin, Dave Scott, Mike Parsons, Debbie Smith, Liz Bussey, Seija Usitalo of Farmington, Michigan, Benjamin Gilliand of Bell Helicopter Textron, Gerard Martowlis, Scott Martelle, Larry Smith, Bill Houghtaling, Dan Schlund, Raymond Li, Yves Rossy, Dan Schlund, and Mac Montandon.

BIBLIOGRAPHY

ABC News. "The Twisted Tale of a Missing Rocket Belt." *Primetime*. October 10, 2002. Retrieved September 1, 2011. www.abcnews.go.com.

Ackerman, Evan. "Welcome to the Future, Here's Your Jetpack." June 21, 2007. Retrieved April 15, 2012. www.ohgizmo.com.

AiResearch Manufacturing Company of Arizona. "Feasibility Study for Small Tactical Aerial Mobility Platform (STAMP)," March 27, 1974. Bernard Lindenbaum Vertical Flight Collection, Wright State University.

Amarena, Carmelo "Nino." Correspondence with the author, 2012.

Amarillo Globe-News. "Lawsuit Seeks to Unravel Mystery of Lost Rocket Belt." July 25, 1999.

"Army Tests One-Man Helicopter." U.S. Army photo via AP Wirephoto, December 29, 1955.

Arrillaga, Pauline. "Where's the Rocket Belt?" Wilmington, NC, *Star-News*, July 27, 1999.

BBC News. "Jetman Africa–Europe Flight Fails." November 25, 2009. Retrieved May 6, 2012. www.jetman.com/history.

BBC News. "London Rocked by Terror Attacks." July 7, 2005. Retrieved April 22, 2012. http://news.bbc.co.uk/2/hi/4659093.stm.

BBC News. "M25 Chaos after Lorry Explosion." August 30, 2005. Retrieved April 20, 2012. www.news .bbc.co.uk.

BBC News. "7/7 Inquests: Coroner Warns over Bomb Ingredient." February 1, 2011. Retrieved April 23, 2012. www.news .bbc.co.uk.

BBC News. "Six Accused of London Bomb Plot." January 15, 2007. Retrieved April 23, 2012. http://news.bbc.co.uk/2/hi/uk_news/6261899.stm.

BBC News. "Three Guilty of Airline Bomb Plot." September 7, 2009. Retrieved April 23, 2012. http://news.bbc.co.uk/2/hi/8242238.stm.

Bell Aerosystems. "Agreement" (contract between Bell and Royal Agricultural & Horticultural Society for Adelaide Fair). 1968, author's collection.
Bell Aerosystems. Contract with William P. Suitor. Dated October 25, 1965, author's collection.
Bell Aerosystems. "Development and Test of the Bell Zero-G Belt." Technical Documentary Report Number AMLR-TDR-63-23. March 1963, author's collection.
Bell Aerosystems. "Individual Lift Device." Contract No. DA23-204-AMC-03712. Reports No. 21.(October 15, 1967) and 35 (December 15, 1968), author's collecion.
Bell Aerosystems. "Light Mobility Systems Missions." Report No. 2203-927001. Undated, author's collection.
Bell Aerosystems. "Proposal for a Jet Flying Belt." Report No. D2203-953001. May 1964, author's collection.
Bell Aerosystems Company. "Small rocket lift device." Phase I (Design, fabrication and static testing). March 1961, author's collection.
Bell Aerosystems Company. "Small Rocket Lift Device." Phase II (Testing of the assembled unit). July 1961, author's collection.
Bell Aerosystems Company. "Small Rocket Lift Device." Phase III (Continued testing). April 1963, author's collection.
Bengsston, Erik. Correspondence with the author, May 2012.
Bernard Lindenbaum Vertical Flight Collection. Wright State University.
Bohr, A. H. "Jet Lift Concepts to Improve Individual Soldier Mobility," June 3, 1959. Courtesy of Bernard Lindenbaum Vertical Flight Collection. Wright State University.
Bohr, A. H., "Rocket JumpBelt: A Proposed Experimental Program," August 1, 1960, Bernard Lindenbaum Vertical Flight Collection. Wright State University.
Boyle, Alan. "Is This Your Jetpack?" *Cosmic Log*. Retrieved on September 10, 2012. http://cosmiclog.nbcnews.com/_news/2008/07/29/4351508-is-this-your-jetpack?lite.
Brown, Douglas. "Wanting to Jet into the Future." *Denver Post*. July 17, 2007.
Brown, Paul. *The Rocketbelt Caper*. Newcastle upon Tyne, UK: Superelastic, 2009.
Buffalo News. "Gordon R. Yaeger, Piloted Rocket Belt." Obituary. January 25, 2005.
Buffalo News. "Harold M. Graham, Pioneering Flier with Bell Rocket Belt." Obituary. October 28, 2009.
Bulaga, Robert. Interviews with author, April 2012.
Burge, David. "The Edison of Crazy." *Garage*. Vol. 16, 58.
Burgess, Phil. "Rocket Roundup." Retrieved April 22, 2012. www.nhra.com.
CBS. *Late Show with David Letterman*. May 4, 2012.
Chalberg, Dru. "Teenager Is One of Rocket Belt Flyers." *Sacramento Bee*. September 10, 1964.
Chicago Daily News. June 27, 1963.

Clark, Jerome. *Unexplained: Strange Sightings, Incredible Occurrences & Puzzling Physical Phenomena.* Detroit: Visible Ink Press, 1999.

CNN. "Personal 'Jetpack' Gets off the Ground." CNN Tech. February 6, 2002. Retrieved on April 10, 2012. www.articles.cnn.com.

Code of Federal Regulations. Title 14: Aeronautics and Space. Sec. 21.191 "Experimental certificates."

Coppola, Lee. "Capturing Flight on Film." *Buffalo News Magazine.* August 3, 1983.

Courter, Robert. "I Fly the Man Rockets." Ed. James Joseph. *Popular Mechanics.* October 1964.

Courter, Robert. Interviews with the author, August 2012.

Courter, Robert. "What It's Like to Fly the New Jet Belt." *Popular Science.* November 1969.

Czaplyski, Vincent. "Oldies & Oddities: Son of Rocket Belt." *Air & Space.* July 2002.

Daily Press (Newport News). "Selmer A. Sundby." Obituary. January 1, 2004.

Discovery News. "'Jetman' Soars over Rio de Janeiro: Big Vid." May 7, 2012. Retrieved May 7, 2012. www.news.discovery.com.

DuBarry, John. "The Sky-High Invention." *True.* September 1956.

Flight. "Flying Platform." November 2,1956.

Flight. "Rotor and Ring." April 22, 1955.

Flight. "VZ-1E Flying Platform." March 1958.

Flight International. "Bell's Jet Belt." July 11, 1968.

Flight International. "F107 (WR-19)." January 2, 1975.

Flight International. "Late News and Amendments." June 6, 1963.

Flight International. "Outdoing Jules Verne." June 20, 1963.

Flight International. "Paris Report." June 13, 1963.

Flight International. "Sud Ludion." May 23, 1968.

Flight International. "Turbine Engines of the World." January 4, 1973.

Flight International. "Williams Research." January 7, 1971.

Fonseca, Felicia. "Swiss Jetman Cancels Grand Canyon Flight." *Deseret News.* May 6, 2011. Retrieved May 6, 2012. www.deseretnews.com/article/700133235 /Swiss-JetMan-cancels-Grand-Canyon-flight.html.

Frank, Thomas. "TSA: Airline Ban on Liquids Won't Be Lifted Soon." September 9, 2009. Retrieved April 28, 2012. www.usatoday.com.

Frauenfelder, Mark. "Rocketbelt Convention." August 2, 2007. Retrieved April 20, 2012. www.boingboing.net.

Gannon, Robert. "This Man Can Broad-Jump 368 Feet." *Popular Science.* December 1961.

Gibson-Frusher, Sheri. Letter to Shari Sheffield. September 1, 2004. www.uspto.gov.

Gibson, Howard McKinnie, Jr. "ROCKETBELT," Trademark Registration No. 3,003,697, filed January 16, 2004; registered October 4, 2005; cancelled October 5, 2012.

Gibson, Howard McKinnie, Jr. "ROCKETMAN," Trademark Registration No. 3,579,081, filed March 25, 2008; registered February 24, 2009.

Gijsberts, Peter. "First International Rocketbelt Convention." Press release, August 31, 2006.

Gijsberts, Peter. Interviews with the author, 2011–2012.

Glaister, Dan. "First 'Practical Jetpack' Clears for Take-Off." *Guardian*. July 29, 2008.

Haggerty, James J., Jr., and Cornelius Ryan. "The Navy Comes Up with a Real 'Flying Saucer.'" *Collier's*. April 29, 1955.

Hambling, David. "Is It a Bird? Is It a Plane?" *Guardian*. September 13, 2000.

Hargrove, Brantley. "Hero Pilot Hal Graham's Hard Fall to Earth." *Nashville Scene*. November 26, 2009.

Hiller Aviation Museum. "Flying Platform." Retrieved October 31, 2012. www.hiller.org/flying-platform.shtml.

Hiller Helicopters. "Stability Analyses of Flying Platform in Hovering and Forward Flight." Advanced Research Division. October 12, 1956, author's collection.

Houghtaling, Bill. Interview with the author, May 2012.

Hulbert, John K. and Moore, Wendell, F. "Personnel Propulsion Unit." US Patent 3,243,144. Filed July 17, 1964; issued March 29, 1966.

Jenkins, Dennis R., Tony Landis, and Jay Miller. *American X-Vehicles: An Inventory-X-1 to X-50*. Monographs in Aerospace History. Vol. 31, National Aeronautics and Space Administration, Washington, DC, June 2003.

Kocivar, Ben. "Turbo-Fan Powered Flying Carpet." *Popular Science*. September 1982.

Lee, James. "Start." *Wired*. December 2006.

LeGrande, Ray, Sr. Interview with the author, March 2012.

Leyes, Richard A., and William A. Fleming. *The History of North American Small Gas Turbine Aircraft Engines*. Reston, VA: American Institute of Aeronautics and Astronautics, Inc., 1999.

Li, Raymond. Interview with the author, 2012.

Li, Raymond, "Personal Propulsion Device." US Patent 7,258,301. Filed March 23, 2005; issued August 21, 2007.

Life. "New Army Drill; 5-4-3-2-1-At-ten-tion!" October 20, 1961.

Lindenbaum, Bernard. "Developed Aircraft, A Historic Report." Vol. I (1940–1986), US Air Force and US Navy sponsored report, issued June 26, 1986, author's collection.

Lozano, Juan Manuel. Interview with the author, April and May 2012.

MAD. "Amazing Military Rocket-Belt Developed: Army Unsure of Practical Use." December 1961.

Mail Online. "Pictured: Rocketman Flies over Alps with Jet-Pack Strapped to His Back." May 15, 2008. Retrieved May 6, 2012. www.dailymail.co.uk.

Marin, Marcel. "Steering of Portable Reaction Motors." US Patent 2,509,603. Filed February 24, 1945; issued May 30, 1950.

Martin, Glenn Neil. "Propulsion Device." US Patent 7,484,687. Filed October 26, 2005; issued February 3, 2009.

Martowlis, Gerard. Interview with the author, May 2012.

"Memorandum Opinion & Order Granting Defendant's Motion for Summary Judgment." Powerhouse Productions, Inc. et al v Troy Widgery et al, No. 4:07-cv-071, US District Court (E. D. Tex.)

Metzgar, T. L. *WASP.* Somerset, PA: Deeter Gap Publishing, 1987.

Michaelson, Ky. Interview with the author, April, 2012.

Miele, Frank. "Giving the Devil His Due." *Skeptic.* Vol. 2, No. 4, 58–61.

Mone, Gregory. "The DIY Flier." *Popular Science.* December 2008.

Montandon, Mac. *Jetpack Dreams: One Man's Up and Down (But Mostly Down) Search for the Greatest Invention that Never Was.* New York: Da Capo Press, 2008.

Moore, Thomas M. Lockwood Airfoil Used in Conjunction with Man Transport Device. US Patent 4,040,577. Filed January 17, 1977; issued August 9, 1977.

Moore, Wendell F. "Propulsion Unit." US Patent 3,021,095. Filed June 10, 1960; issued June 10, 1960.

Mosier, Michael. "Single Passenger Aircraft." US Patent 6,488,232, filed December 16, 1998; issued December 3, 2002.

MSNBC. "JetMan Pulls Off Grand Canyon Flight—Quietly." May 10, 2011. Retrieved May 6, 2012. www.msnbc.msn.com/id/42977241/ns/us_news-life/t/jetman-pulls-grand-canyon-flight-quietly/#.UJE_vYbZ1ro.

Multi-Laker. "Williams Research Chief Will Be Speaker at February Meet." Vol. 21, No. 2, February 20, 1974.

Neracher, Arnold. Correspondence with the author. Translated by Jill Warner Dailey. April 2012.

New York Times. "Frank R. Paul Dead; Illustrator Was 79." Obituary. June 30, 1963.

The Official Newsletter for the Martin Jetpack. "Taking Off." February 2011. www.martinjetpack.com.

Olney, Ross S. "America's Aerial Foot Soldiers." *Science and Mechanics.* March 1963.

Popular Science. "Jet Pack Turns Astronaut into Human Spacecraft." November 1962.

Powerhouse Productions. *Introducing the Rocketman.* Brochure, undated.

Purdy, Kevin. "Airshow: Rocket Men Unite." Medina, NY, *Journal-Register,* August 11, 2007.

Ravilious, Kate. "'Jet Man' Crosses English Channel Like a Human Rocket." Retrieved May 7, 2012. www.nationalgeographic.com.

Roach, Robert. "The First Rocket Belt." *Technology and Culture* Vol. 4, no. 4, Autumn 1963: 490–98.

Roach, Robert. Interview with the author, April 2012.

Robertson, Arthur C., and Stuart, Joseph, III. "Vertical Take-Off Flying Platform." US Patent 2,953,321. Filed February 27, 1956; issued September 20, 1960.

Rogers, Mike. *VTOL Military Research Aircraft*. New York: Orion Books, 1989.
Rossy, Yves. Correspondence with the author, April, August 2012.
Rynin, N. A. *Interplanetary Flight and Communication*. Vol. 2, No. 4. 1929. Israel Program for Scientific Translations, Jerusalem, 1971.
Saturday Review. "Man Learns to Fly in a Steam-Powered Corset." July 7, 1961.
Schlund, Dan. Interview with author, May 2012.
Schmeck, Harold M., Jr. "Jet Flying Belt Is Devised to Carry Man for Miles." *New York Times*. June 28, 1968.
Smith, Larry. Interview with the author, July 2012.
Smith, Larry. "Rocket Men." *Slate*. October 4, 2006. Retrieved April 27, 2012. www.slate.com.
Smithsonian National Air and Space Museum. "Rocket Belt." Retrieved January 23, 2012. www.nasm.si.edu.
Sofge, Erik. "The Inside Story of When Jet Packs Really Are Coming." *Popular Mechanics*. October 2009.
Spenser, Jay. *Vertical Challenge: The Hiller Aircraft Story*. Seattle: University of Washington Press, 2003.
Spencer, John. Interview with the author, April 2012.
Suitor, William. Interviews with author, April–August 2012.
Suitor, William. *Rocketbelt Pilot's Manual*. Burlington, Ontario: Apogee Books, 2009.
Taylor, Michael. *The World's Strangest Aircraft*. New York: Metrobooks, 2004.
Tech.co.uk. "Ditch the Car and Fly to Work with a Jet-Pack." Techradar, posted December 14, 2007. Retrieved April 15, 2012. www.techradar.com.
Telegraph (Nashua, New Hampshire). "Army Testing One-Man Flying Platform." June 10, 1982.
Thunderbolt Aerosystems Inc. "Thunderbolt Aerosystems Inc. Unveils Thunderpack," press release. January 24, 2008. Retrieved April 9, 2012. www.thunderbelt.net.
Times Daily (Alabama). "Rocket Belt Designer Wins Lawsuit." July 28, 1999.
Tozer, Eliot. "Man's First Leap toward Free Flight." *Popular Science*. December 1958.
Transportation Security Administration. "Make Your Trip Better Using 3-1-1." Retrieved April 23, 2012. www.tsa.gov/traveler-information/make-your-trip-better-using-3-1-1.
US Army. "The de Lackner Aerocycle—An Early Flying Platform." Retrieved April 4, 2012. www.transportation.army.mil/museum/transportation%20museum/delackner.htm.
USA Today. "Rocket Belt Rights Returned." November 23, 1999.
Verne, Jules. *The Clipper of the Clouds*. London, 1887.
Voss, Mark. Interview with the author, April 2012.
Williams, Sam B. "Twin Spool Gas Turbine Engine with Axial and Centrifugal Compressors." US Patent 3,357,176. Filed September 22, 1965; issued December 12, 1967.

Wahl, Paul. "Jet Flight without Wings," *Popular Science*. April 1974.

Watertown Daily Times. "Wow, What a Lineup!" August 11, 1969.

Wettergren, Maria and Johannes Cleris. "Explosion i Aspereds industriomrade." GP, July 23, 2010. Retrieved on April 20, 2012. Translated by Markku Jaakkola and Seija Usitalo. www.gp.se.

Widgery, Troy. Interviews with the author, April–August 2012.

Williams Research Corporation. "System Specification for WASP II." September 24, 1980.

Williams, Sam B. Airborne Vehicle. US Patent 4,447,024. Filed February 8, 1982; issued May 8, 1984.

Williams, Sam B. Flight belt. US Patent 3,443,775. Filed June 23, 1965; issued May 13, 1969.

Wolf, Jamie. "Canceled Flight." *New York Times Magazine*. June 11, 2000.

Zimmerman, Charles H. Helicopter Flying Apparatus. US Patent 2,417,896. Filed August 10, 1943; issued March 25, 1947.

NOTES

INTRODUCTION

1. Jules Verne, *The Clipper of the Clouds* (1887), p. 31. The book is also known by the title *Robur the Conqueror*.
2. It is often said that the character on the cover of the magazine was Buck Rogers, but that is not the case. Buck Rogers appeared in a different comic within the same issue. See Mac Montandon, *Jetpack Dreams: One Man's Up and Down (But Mostly Down) Search for the Greatest Invention That Never Was* (New York: Da Capo Press, 2008), 16.
3. Frank R. Paul died in 1963 and is best known for his futuristic illustrations of science fiction magazines. See "Frank R. Paul Dead; Illustrator was 79," *New York Times* obituary, June 30, 1963.
4. I am aware that throughout this book I refer to *men* who flew jet packs. In this case it is true that all of the pilots of the individual lift devices were men until quite late in the story. In that instance I will obviously point it out. It did not make sense to me to encumber the story with more generic pronouns when they were not necessary or accurate.

CHAPTER I: FLYING SHOES AND HOVERING PLATFORMS

1. John DuBarry, "The Sky-High Invention," *True*, September 1956, 46–48, 103–5.
2. James J. Haggerty Jr. and Cornelius Ryan, "The Navy Comes Up with a Real 'Flying Saucer,'" *Collier's*, April 29, 1955, 34.
3. Jay P. Spenser, *Vertical Challenge: The Hiller Aircraft Story* (Seattle: 1st Book Library, 2003), viii.
4. Charles H. Zimmerman, Helicopter Flying Apparatus, US Patent 2,417,896, filed August 10, 1943; issued March 25, 1947.
5. Spenser, 112–3.

6. One source says it was 1947. DuBarry, 104.
7. Haggerty, 34. It is unclear if Hiller hired Zimmerman or merely cut a deal to examine and test the Flying Shoes. See DuBarry, 104.
8. Spenser, 113.
9. Haggerty, 34.
10. DuBarry, 104.
11. Ibid.

CHAPTER 2: THE HILLER AND DE LACKNER FLYING PLATFORMS

1. Spenser, 113.
2. DuBarry, 105.
3. Spenser, 113.
4. DuBarry, 105.
5. Haggerty, 35.
6. DuBarry, 105.
7. Spenser, 114.
8. "Rotor and Ring," *Flight*, April 22, 1955, 510.
9. *Flight,* 1955, 510.
10. Haggerty, 30.
11. Haggerty, 32–33.
12. Haggerty, 32.
13. Ibid, 33.
14. Ibid, 33.
15. "Lift degradation" is from Lindenbaum, 125.
16. Haggerty, 35.
17. Ibid, 35.
18. "Flying Platform," *Flight*, November 1956, 725.
19. "VZ-1E Flying Platform," *Flight*, March 1958, 396.
20. Ibid.
21. Spenser, 116.
22. Hiller Helicopters, Advanced Research Division, "Stability Analyses of Flying Platform in Hovering and Forward Flight," October 12, 1956. Hereafter, "Hiller."
23. Hiller, 18.
24. Ibid, 33–34.
25. "Flying platform," www.hiller.org/flying-platform.shtml, accessed October 31, 2012. A good description of this unusual airfoil can be found at "Rotor and Ring," *Flight*, April 22, 1955, 510. Hereafter, "*Flight,* 1955."
26. Robertson, Arthur C. and Stuart, Joseph III, "Vertical Take-Off Flying Platform," US Patent 2,953,321, filed February 27, 1956; issued September 20, 1960.
27. DuBarry, 105.

28. "The De Lackner Aerocycle—An Early Flying Platform," retrieved April 4, 2012, www.transchool.lee.army.mil, hereafter, "De Lackner." Another source gives the engine's rating as 44 horsepower. "Army Tests One-Man Helicopter," U.S Army photo via AP Wirephoto December 29, 1955.
29. Mike Rogers, *VTOL Military Research Aircraft* (New York: Orion Books, 1989), 75.
30. "Army Tests One-Man Helicopter," U.S Army photo via AP Wirephoto, December 29, 1955.
31. De Lackner; "Selmer A. Sundby," obituary, *Daily Press* (Newport News), January 1, 2004.
32. DuBarry, 46.
33. De Lackner.

CHAPTER 3: THE AGE OF MAN ROCKETS

1. Eliot Tozer, "Man's First Leap toward Free Flight," *Popular Science*, December 1958, 72. Hereafter, Tozer, "Man's First Leap."
2. Marin, Marcel, "Steering of Portable Reaction Motors," US Patent 2,509,603, filed February 24, 1945; issued May 30, 1950.
3. N. A. Rynin, *Interplanetary Flight and Communication*, Israel Program for Scientific Translations, Jerusalem, (1971), 72, hereafter, "Rynin."
4. Rynin, 74.
5. Montandon, 20–22.
6. Rynin, 1–2.
7. Lindenbaum, 2:41.
8. Tozer, "Man's First Leap," 70–72.
9. Bernard Lindenbaum, *Developed Aircraft: A Historic Report*, vol. 1 *1940–1986*, US Air Force and US Navy sponsored report, issued June 26, 1986, 2:35. Hereafter, "Lindenbaum."
10. Tozer, "Man's First Leap," 71.
11. Ibid, 73.
12. Lindenbaum, 2:11.

CHAPTER 4: THE QUEST FOR THE ROCKET BELT

1. Lindenbaum, 2:11.
2. Lindenbaum, 2:11. He notes that Bell had come up with the idea for a flying belt "spontaneously."
3. Lindenbaum, 2:11.
4. Aerojet-General, "Feasibility Study of Small-Rocket Lift Device," Report No. 1751, February 1960, 4. Bernard Lindenbaum Vertical Flight Collection, Wright State University. Hereafter, "Aerojet."

5. Aerojet, 5.
6. Montandon, quoting an interview with William Suitor, 78.
7. Aerojet, 12.
8. Ibid, 13.
9. Ibid, 5.
10. Ibid, 1.
11. Lindenbaum, 2:29.
12. A. H. Bohr, Reaction Motors Division Thiokol Chemical Corporation, "Jet Lift Concepts to Improve Individual Soldier Mobility, TR-737," June 3, 1959, Bernard Lindenbaum Vertical Flight Collection, Wright State University. Hereafter, "Bohr, 1959."
13. Bohr, 1959, 15.
14. Ibid, 7.
15. Ibid, 3.
16. Ibid, 4.
17. Ibid, 1.
18. Ibid, 8.
19. A. H. Bohr, "Rocket JumpBelt: A Proposed Experimental Program," August 1,1960, Bernard Lindenbaum Vertical Flight Collection, Wright State University. Hereafter, "Bohr, 1960." Frontispiece.
20. Bohr, 1960, 1.
21. Ibid, 4.
22. Ibid, figure 13.
23. Ibid, figure 19.
24. Ibid, 58–59.
25. William P. Suitor, *Rocketbelt Pilot's Manual* (Burlington, Ontario: Apogee Books, 2009), 11.
26. Montandon gives the figure of $25,000, 33.
27. Bell Aerosystems Company, "Small Rocket Lift Device," Phase I (Design, Fabrication and Static Testing), March 1961; hereafter, "Phase I."

CHAPTER 5: WENDELL F. MOORE AND THE BELL ROCKET BELT

1. Bell Aerosystems Company, "Proposal for a Jet Flying Belt," Report No. D2203-953001, May 1964, C-12; hereafter "Proposal 1964."
2. Dennis R. Jenkins, Tony Landis, and Jay Miller, *American X-Vehicles: An Inventory X-1 to X-50,* no. 31, Monographs in Aerospace History, National Aeronautics and Space Administration, Washington, DC, June 2003, 5.
3. Bell Aerosystems, "Development and Test of the Bell Zero-G Belt," Technical Documentary Report Number AMLR-TDR-63-23, March 1963, 12.
4. Montandon, 31.

5. 1953 is from Lindenbaum, 2:12, but is found elsewhere. See Robert D. Roach Jr., "The First Rocket Belt," *Technology and Culture* (Vol. 4, no. 4, Autumn 1963): 490–498. Hereafter, "Roach." Also, Bell Aerosystems, "Development and Test of the Bell Zero-G Belt," Technical Documentary Report Number AMLR-TDR-63-23, March 1963, 12.
6. Roach, 491. Also, Paul Brown, *The Rocketbelt Caper,* (Newcastle upon Tyne, UK: Superelastic, 2009), 20–21. Hereafter, "Brown." The story is also recounted in Montandon, 29, although it could be that Moore merely explained this more than once.
7. Robert Gannon, "This Man Can Broad-Jump 368 Feet," *Popular Science*, December 1961, 106. Hereafter, Gannon, "Broad-Jump."
8. Montandon, 31.
9. Roach says the first tethered test was run in the "winter of 1958," and in that respect, he is probably wrong, but not by much. See Roach, 492. Others in Bell also gave 1958 as the date the program began. See Bell Aerosystems, "Development and Test of the Bell Zero-G Belt," Technical Documentary Report Number AMLR-TDR-63-23, March 1963, 17. Hereafter, "Zero-G."
10. Bell Aerosystems, "Small Rocket Lift Device," Phase II (Testing of the Assembled Unit), July 1961, ii. Hereafter, "Phase II."
11. Roach interview, April 2012. Robert Roach worked at Bell for twenty-eight years. Before that, he worked for the US Army at their arsenal where he worked on all manner of things being fired by explosives, as small as bullets to as large as nuclear-tipped projectiles designed to be fired from sixteen-inch naval guns.
12. Roach, 493.
13. Lee Coppola, "Capturing Flight on Film," *Buffalo News Magazine*, August 3, 1983.
14. Montandon, quoting an interview with William Suitor, 78.
15. Phase I.
16. Phase I, 33. In other places, the name is given as National Waterlift Company. I am writing it here as it appears in the Bell report, which is correct. According to the corporate filings in Michigan, "National Water Lift Company" was founded in 1956 and retained the name until 1986, when they changed their name to "NWL Control Systems, Inc." The company's name derived from its original business of selling devices to lift water from cisterns whereby people could pump cistern water, which was soft, to do their laundry. The company moved into making hydraulic parts for airplanes during World War II. Bob Kutsche, interview with author, May 2012.
17. Roach, 494. It was similar in operation to the one in the Mercury program, but was completely re-done for the rocket belt program, according to Kutsche.
18. This might seem improbable but it is not. Machinists routinely work with designs that require precision to the 1/10,000th of an inch. According to

several engineers and machinists interviewed by the author, the next step in measurement is the 1,000,000th. They do not often speak of measurements in the 100,000ths. The measurements in this case very well could have been in the hundreds. That is, for example, something may have called for a tolerance of 90/1,000,000ths. Obviously, that could be rendered as 9/100,000ths, but it would be an awkward and uncustomary way of expressing it. In this case, one belt builder told the author that the valve in his belt had tolerances in the 200/1,000,000ths range.

19. Bob Kutsche, interviewed by author, 2012. For example of language on valve, see rocket belt at University of Buffalo, photos included in this work.
20. Phase I, 38.
21. Lindenbaum, 2:22.
22. Suitor, *Manual*, 27.
23. Lindenbaum, for one, 2:30. He also mentions the earplugs.
24. Phase I, 9.
25. Ibid, 6.
26. Phase II, 1.
27. Phase I, 19.
28. Phase II, 3.
29. Roach, 494.
30. Phase I, 92.
31. Ibid, 13.
32. Phase II, 8.
33. Phase I, 13.
34. Ibid, 18.
35. Suitor, *Manual*, 6. This page is a reprint of a Bell document describing the rocket belt.
36. Phase I, 21.
37. Ibid, 23.
38. Roach, interview.
39. Phase I, 36 for the number of eighty-eight firings.
40. Phase I, 5.
41. Ibid, 62.
42. Phase II, 13.
43. Phase I, 54.
44. Ibid, 1.
45. Phase II, 136.
46. Ibid, 8.
47. Phase II, 45. Roach suggests that this hangar was the same as the one used originally, but the Phase II reports indicate that it was a different, larger hangar. See Roach, 494.

48. Phase II, 8.
49. Ibid, 12.
50. Montandon, 34.
51. Phase II, 64-65.
52. Phase II, 69. Roach says it was "approximately" ten feet but the eight-foot figure seems more accurate. It is from the report of the event while Roach was working from interviews a few years later. See Roach, 495.
53. Phase II, 12.

CHAPTER 6: THE ROCKET BELT FLIES

1. Montandon, 35. That their desks were adjacent is from an interview of Bob Roach by the author, April 2012.
2. This story is told many places. One such also mentions that Graham had worked on the rocket belt previously. See Roach, 495.
3. "Man Learns to Fly in a Steam-Powered Corset," *Saturday Review*, July 7, 1961, 29. Hereafter, "*Sat. Rev.*"
4. Brantley Hargrove, "Hero Pilot Hal Graham's Hard Fall to Earth," *Nashville Scene*, November 26, 2009. Hereafter, "Hargrove."
5. John Spencer interviewed by the author, April 2012.
6. Phase II, 70.
7. Ibid, 54.
8. Ibid, 40.
9. Ibid, 12.
10. Phase II, 89. The details of this day are also described in Roach, 490–498.
11. Roach, 495.
12. Phase II, 14. Interestingly, this report suggests the flight was exactly 100 feet long. All other sources say it was 112 feet.
13. Suitor, *Manual,* 11–12. "Harold M. Graham, Pioneering Flier with Bell Rocket Belt," obituary, *Buffalo News*, October 28, 2009. Hereafter, "Harold Graham obituary."
14. Roach, 495.
15. Ibid, 496.
16. Phase II, 46.
17. Ibid, 89.
18. Ibid, 45.
19. Ibid, 56.
20. Ibid, 56.
21. Ibid, 18-19.
22. Ibid, 20.
23. Ibid, 26.

24. Hargrove.
25. Phase II, 48.
26. Ibid, 48.
27. Ibid, 123–124.
28. Ibid, 127.
29. Ibid.
30. Ibid, 4.
31. "Portable Army Rocket Propels Man 150 Feet in 14-Second Test Flight," *New York Times*, June 9, 1961.
32. Roach, interview.
33. Hargrove.
34. Roach, 496. Another source gives the attendance figure as being six thousand, Montandon, 39.
35. Hargrove.
36. *Sat. Rev.*
37. Ibid.
38. *MAD*, December 1961, 4–7.
39. "New Army Drill; 5-4-3-2-1-At-ten-tion!" *Life*, October 20, 1961. Hereafter, "*Life*."
40. Roach, 497.
41. Brown, 27.
42. Roach, 496–7.
43. *PS*, Dec. 1961, 104–106, 192.
44. *PS*, Dec. 1961, 192.
45. Ibid.
46. Hargrove.
47. Ibid.
48. Bell Aerosystems, "Small Rocket Lift Device," Phase III (Continued Testing), April 1963; hereafter, "Phase III."
49. Phase III, 1–2.
50. Ibid, 25.
51. Ibid, 41.
52. Ibid, 43.
53. Ibid, 45.
54. Ibid, 48-49.
55. Brown wrote that Kedzierski was seventeen when he was hired. This is contradicted by various Bell reports and it seems unlikely that Bell would hire a minor for such a dangerous job. See Brown, 28.
56. Phase III, 142.
57. Ibid, 155.
58. Ibid, 170.
59. Lindenbaum, 2:18.

60. Proposal 1964, C-12.
61. "Jet Pack Turns Astronaut into Human Spacecraft," *Popular Science*, November 1962, 85. Hereafter, "Jet Pack Turns Astronaut."

CHAPTER 7: THE BELL ROCKET MEN

1. Lindenbaum, 2:18.
2. Roach, interview.
3. Brantley Hargrove, "Hero Pilot Hal Graham's Hard Fall to Earth," *Nashville Scene*, November 26, 2009, and Brown, 27. Hal Graham told the story the first time publicly in 2006 at the rocket belt convention, which explains why it was not reported previously. Also see Larry Smith, "Rocket Men: The Guys Who Really, Really Want to Fly," www.slate.com/articles/news_and_politics/dispatches/2006/10/rocket_men.html, accessed October 31, 2012.
4. Roach, 497. Roach knew about Graham's crash. During an interview in 2012 he mentioned it but pointed out that the crash was most likely caused by the unstable platform that Graham was on and not attributable to the rocket belt itself.
5. Roach, 497.
6. Brown, 36.
7. Bob Courter, interviewed by the author, August 2012.
8. Proposal 1964, C-13. The proposal gives the number of 140 hours for Courter's combat; a newsletter called *The Multi-Laker* gives it as 450. "Williams Research Chief Will Be Speaker at February Meet." *The Multi-Laker*. February 20, 1974, 4.
9. Brown, 32.
10. Brown, 32. This figure seems high when compared to the $10,000 paid six years later for the team to go to Australia.
11. *Chicago Daily News*, June 27, 1963.
12. Montandon, 40.
13. Ross S. Olney, "America's Aerial Foot Soldiers," *Science and Mechanics*, March 1963, 42–45. Hereafter, "*SM*, Mar. 1963."
14. *SM*, Mar. 1963, 45.
15. "Late News and Amendments," *Flight International*, June 6, 1963, 888.
16. Ibid.
17. Courter, interview.
18. "Paris Report," *Flight International*, June 13, 1963, 906.
19. "Outdoing Jules Verne," *Flight International*, June 20, 1963, 954, hereafter, "*Flight* 1963."
20. *Flight* 1963, 957.
21. Robert F. Courter, Jr., as told to James Joseph, "I Fly the Man Rockets," *Popular Mechanics*, October 1964, 120. Hereafter, "*PM*, Oct. 1964."

22. Montandon, 45.
23. Roach, 497.
24. *PM*, Oct. 1964, 122.
25. Ibid, 123, 212.
26. Suitor, *Manual*, 23.
27. Ibid, 59.
28. Ibid, 55.
29. Ibid, 60.
30. *PM*, Oct. 1964, 123.

CHAPTER 8: THE SUD LUDION AND THE POGO

1. Lindenbaum, 2:53. Lindenbaum suggests the name came from the tanks on the back of the device looking like those on the back of someone wearing tanks for breathing underwater.
2. Lindenbaum, 2:60.
3. Ibid, 2:54–55.
4. Ibid, 2:58–59.
5. "Sud Ludion," *Flight International*, May 23, 1968, 797. The article actually says the payload figure is 661 pounds but that appears to be a typo. The entire unit fully loaded with fuel weighed less than 400 pounds. See Lindenbaum, 2:55.
6. The flight information is taken from a Bell promotional film that can be found on YouTube by searching the phrase "flying seat." Retrieved April 12, 2012.
7. Courter, interview.
8. Lindenbaum, 2:16.
9. Ibid, 2:28.
10. Lindenbaum, 2:28. Also, "Pogo," *Flight International*, February 5, 1970, 200.
11. Roach mentions the surface transport in his 1963 *Technology and Culture* article, 498, and he mentioned the lunar lander in an interview with the author.
12. John Spencer, correspondence with the author, April 2012.
13. Zero-G, iii.
14. Ibid, 8.
15. Ibid, 61.
16. Popular Science, "Jet Pack Turns Astronaut," 85.
17. Wendell F. Moore, "Personnel Flying Device," US Patent 3,381,917, filed November 8, 1966 and issued May 7, 1968.
18. Lindenbaum, 3:26.

CHAPTER 9: WILLIAM P. SUITOR: ROCKET MAN

1. Suitor, *Manual*, inside back flap. In Brown, the question is phrased, "Hey kid, you want a job?" Brown, 30.

2. Brown, 30–31.
3. Phase III.
4. Brown, 31.
5. Ibid, 32.
6. Bell Aerosystems Company, contract with William P. Suitor, dated October 25, 1965, hereafter "Suitor, 1965 Contract."
7. Suitor, 1965 Contract, 2.
8. Ibid, 6.
9. Dru Chalberg, "Teenager Is One of Rocket Belt Flyers," *Sacramento Bee*, September 10, 1964, C8, hereafter, "Chalberg."
10. Chalberg.
11. Brown, 34.
12. Ibid.
13. Ibid, 35.
14. Suitor, *Manual,* inside back flap.
15. Courter, "New Jet Belt," 59.
16. Brown, 32. For more on the world's fair and the Chrysler display, see Steve Lehto, *Chrysler's Turbine Car: The Rise and Fall of Detroit's Coolest Creation* (Chicago: Chicago Review Press, 2010), 86–87.
17. Courter, "New Jet Belt," 59.
18. Bell Aerosystems Company, "Agreement" (contract between Bell and Royal Agricultural & Horticultural Society for Adelaide Fair) 1968. Hereafter, "Bell-Royal contract."
19. Bell-Royal contract.
20. T. L. Metzgar, *WASP* (Somerset, PA: Deeter Gap Publishing, 1987), 18; hereafter, "Metzgar."
21. Brown, 36.
22. Suitor, *Manual,* 13.
23. *Watertown Daily Times*, August 11, 1969, 7. This is a full-page ad. Fair attendance was 550,000 according to Sharron Pearson, a New York state employee affiliated with the New York state fair, in correspondence with the author, April 2012.
24. Suitor interview with author April–August 2012.
25. *Buffalo News*, "Gordon R. Yaeger, Piloted Rocket Belt," obituary, January 25, 2005.

CHAPTER 10: THE JET FLYING BELT

1. Roach, interview. "The short duration characteristic" was also cited by Lindenbaum as the primary drawback seen by the military with respect to the rocket-powered devices. 2:28.

2. Proposal 1964. The device was known as either the Jet Flying Belt or simply the Jet Belt. For the latter, see "Bell's Jet Belt," *Flight International*, July 11, 1968, 78. Hereafter, *Flight*, 1968.
3. Lindenbaum, 3:20. J. K. Hulbert and Wendell F. Moore, "Personnel Propulsion Unit," US Patent 3,243,144, filed July 17, 1964; issued March 29, 1966.
4. Proposal 1964, E-1. This date is also given in the biography of Sam Williams from the National Aviation Hall of Fame. See www.nationalaviation.org/williams-sam/. Some sources give the date as January 1955.
5. National Aviation Hall of Fame, "Sam Williams," from the website www.nationalaviation.org.
6. Proposal 1964, E-9. See Sam B. Williams, "Twin spool gas turbine engine with axial and centrifugal compressors," US Patent 3,357,176, filed September 22, 1965; issued December 12, 1967.
7. Richard A. Leyes and William A. Fleming, "The History of North American Small Gas Turbine Aircraft Engines," American Institute of Aeronautics and Astronautics, Inc., Reston, VA:1999: 399. Hereafter, "Leyes and Fleming."
8. Proposal 1964, E-6.
9. Ibid, I-1.
10. Ibid, II-1.
11. Ibid, introductory cover letter, 3–4.
12. *Flight* 1968, 78.
13. Proposal 1964, kerosene on III-1, thrust figures from III-2, weight of the engine at III-6.
14. Proposal 1964, III–4.
15. Ibid, III–14.
16. Bell Aerosystems, "Individual Lift Device," Contract No. DA23-204-AMC-03712, Reports No. 21, (October 15, 1967), 4. Hereafter, "ILD, No. 21."
17. Sam B. Williams, "Twin spool gas turbine engine with axial and centrifugal compressors," US Patent 3,357,176, filed September 22, 1965; issued December 12, 1967.
18. ILD, No. 21, 4.
19. Ibid.
20. *Flight* 1968, 78.
21. Lindenbaum, 3:17.
22. ILD, No. 21, 4–5.
23. Spencer, interview, April 2012.
24. ILD, No. 21, 4
25. Lindenbaum, 3:17–18.
26. Proposal 1964, III–14; photo of woman is figure III–15.
27. Proposal 1964, III–33.
28. Ibid, III–37.

29. *Flight* 1968, 78.
30. Spencer, interview.
31. ILD, No. 21, 1.
32. Ibid, 2.
33. Ibid, 37.
34. Spencer, interview.
35. The same figures of three thousand flights with 100 percent reliability were given in *Flight* 1968, 78.
36. Harold M. Schmeck, Jr., "Jet Flying Belt Is Devised to Carry Man for Miles," *New York Times*, June 28, 1968.
37. *Flight* 1968, 78. Another source gives the date as March 1969. See Leyes and Fleming, 403.
38. Sam B. Williams, "Flight belt," US Patent 3,443,775, filed June 23, 1965; issued May 13, 1969.
39. ILD, No. 21, 33.
40. Spencer, interview.
41. Ibid.
42. Bell Aerosystems, "Light Mobility Systems Missions," Report No. 2203-927001, no date, 1. Hereafter, "LMS." This report is not dated but at one point it states that the jet belt "is scheduled to fly in April 1968," which suggests that the report was issued before that date.
43. LMS, 5.
44. Ibid, 7.
45. Ibid, 38–39.
46. The date and the location of the picture are from www.corbisimages.com/stock-photo/rights-managed/DC004453/human-jet-pack-designer-with-test-pilot?popup=1; some details of the day are from Suitor, correspondence with the author, April–August 2012.
47. Robert Courter, "What It's Like to Fly the New Jet Belt," *Popular Science*, November 1969, 55. Hereafter, Courter, "New Jet Belt."
48. Courter, "New Jet Belt," 55–56.
49. Ibid, 56.
50. Ibid, 57.
51. Ibid, 57–58.
52. Spencer, interview.
53. Lindenbaum, 3:17.
54. Spencer, interview.
55. Montandon, 47.
56. Spencer, interview.
57. Brown, 37.
58. Montandon, 35.

59. Roach interview.
60. Spencer, interview.
61. Ibid.
62. Lindenbaum, 3:20, the price is from Metzgar, 21.
63. "Williams Research," *Flight International*, January 7, 1971, 28.
64. Brown, 38.
65. Proposal, C-13.
66. Spencer, interview.
67. "Williams Research," *Flight International*, January 7, 1971, 28.
68. Lindenbaum, 3:29.
69. "Lockwood Airfoil Used in Conjunction with Man Transport Device," U.S. Patent 4,040,577, granted August 9, 1977.

CHAPTER 11: THE WASP

1. Lindenbaum, 3:32.
2. Ibid, 3:37.
3. Ibid, 3:39.
4. Ibid, 3:38.
5. Ibid, 3:37.
6. *Flight International*, January 2, 1975, 31. Hereafter, "Flight, 1975."
7. Lindenbaum, 3:32.
8. Ibid, 3:41.
9. Metzgar, 23.
10. Metzgar, 24. *Flight* 1975 gives the date as 1973, 31.
11. Lindenbaum, 3:33.
12. Ibid, 3:51.
13. Paul Wahl, "Jet Flight without Wings," *Popular Science*, April 1974, 88–89. Hereafter, Wahl, "Jet Flight." *Flight* 1975 also cites the figure of eleven hundred.
14. Wahl, "Jet Flight," 152.
15. Funding and control of these programs is often a quagmire, regarding who paid for what and under what auspices. Be that as it may, according to the report, the STAMP program was run by the marines and the contract with Garrett was provided by the Naval Weapons Center. AiResearch Manufacturing Company of Arizona, "Feasibility Study for Small Tactical Aerial Mobility Platform (STAMP)," March 27, 1974, Bernard Lindenbaum Vertical Flight Collection, Wright State University, 1. Hereafter, "STAMP."
16. STAMP.
17. Ibid, ii.
18. "Turbine Engines of the World," *Flight International*, January 4, 1973, 28.
19. STAMP, 14.

20. Ibid, 15.
21. Ibid, 73.
22. Lindenbaum, 3:84.
23. STAMP, 45.
24. Ibid, 60.
25. Ibid, 65.
26. Ibid, 80.
27. Ibid, 84.
28. Metzgar, 25.
29. Lindenbaum, 3:89.

CHAPTER 12: THE WASP GETS KINESTHETIC CONTROL: THE WASP II

1. Williams Research Corporation, "System Specification for WASP II," September 24, 1980. Hereafter, "Syst. Spec."
2. Syst. Spec., 38.
3. Ibid, 25.
4. Ibid, 36.
5. Metzgar, 30; Lindenbaum, 3:33.
6. Metzgar, 33.
7. Ibid, 34.
8. Ibid, 34.
9. Lindenbaum, 3:63.
10. Metzgar, 27; Syst. Spec., 20.
11. Lindenbaum, 3:63. Donald L. Underwood, "Preliminary Airworthiness Evaluation (PAE) Williams Aerial Systems Platform II (WASP II) Individual Lift Device," June 1982, 2.
12. Jamie Wolf, "Canceled Flight," *New York Times Magazine*, June 11, 2000.
13. Syst. Spec., 34.
14. Ibid, 37.
15. Metzgar, 40.
16. Ibid.
17. Bulaga, Robert, interview with the author, April 2012. The ballistic parachute in the WASP II was very similar to the one used in the SoloTrek.
18. Metzgar, 40.
19. Ibid.
20. Lindenbaum, 3:34.
21. Metzgar, 41; there is some debate on this. Mark Voss, in a 2012 interview with the author, did not recall any flights longer than twenty minutes. Either way, he said they commonly flew for fifteen to seventeen minutes, which was easily reached.

22. Lindenbaum, 3:63; Underwood, PAE, 8.
23. Metzgar, 41.
24. Ben Kocivar, "Turbo-Fan Powered Flying Carpet," *Popular Science*, September 1982, 67; hereafter Kocivar, "Turbo-Fan Powered."
25. Kocivar, "Turbo-Fan Powered," 67.
26. Metzgar, 43.
27. Ibid, 44.
28. Lindenbaum, 3:34.
29. Lindenbaum, 3:17; Sam B. Williams, "Airborne Vehicle," US Patent 4,447,024, filed February 8, 1982; issued May 8, 1984. Hereafter, "Airborne Vehicle."
30. Airborne Vehicle.
31. Voss, interview.
32. Ibid.
33. Ibid.
34. Ibid.
35. Lindenbaum, 3:34.
36. PAE, 2.
37. Voss, interview.
38. Courter, interview.
39. PAE, 3.
40. Metzgar, 46–47. PAE, Report Documentation Page.
41. PAE, 2.
42. Ibid.
43. Ibid, 5.
44. Metzgar, 53.
45. PAE, 5.
46. Syst. Spec., 11.
47. PAE, 11.
48. Ibid, 5.
49. Ray LeGrande, Sr., interviewed by the author on March 19, 2012. Hereafter, "LeGrande, interview." All references to LeGrande are from this interview unless otherwise noted.
50. Metzgar, 58.
51. Ibid.
52. "Army Testing One-Man Flying Platform," *The Telegraph* (Nashua, New Hampshire), June 10, 1982, 47.
53. While waiting, he saw combat in Desert Storm with the Third Armored Division. He deployed as a platoon sergeant with fifty-eight men and returned with fifty-eight. Along the line he was awarded the Bronze Star. After twenty-one and a half years in the army, he retired to Mecklenburg County, North Carolina, where he spent seventeen years as a deputy sheriff.

54. Lindenbaum, 3:57–58.
55. Ibid, 3:58.
56. Voss, interview.
57. Lindenbaum, 3:58.
58. Ibid, 3:67.
59. Ibid, 3:68.
60. Ibid.
61. Ibid, 3:41.
62. Voss, interview; the WASP II at the Wright Patterson Air Museum is painted as an X-Jet.
63. Voss, interview.
64. Ibid.
65. Wolf, "Canceled Flight."
66. Ibid.

CHAPTER 13: THE ROCKET BELT RETURNS

1. Brown, 39–40.
2. Suitor, *Manual*, 17.
3. Montandon, 50. Some people suggest the Tyler belt was used on *Gilligan's Island* but this is not the case. Tyler's belt did not fly until after Suitor had quit working at Bell; Suitor's last flight as a Bell employee took place over the Labor Day weekend, 1969. *Gilligan's Island* was filmed from 1964 to 1967, two years before the Tyler belt flew. See www.imdb.com.
4. Suitor, *Manual*, 17.
5. Czaplyski, Vincent, "Oldies & Oddities: Son of Rocket Belt," *www.airspacemag.com/flight-today/rocket-belt.html,* accessed October 31, 2012.
6. Brown, 41–42.
7. Ibid, 42–43.
8. That he flew from the top step with Nelson there can be found in several sources. One is Montandon, 182.
9. Brown, 44–45.
10. Montandon, 53.
11. Brown, 46.
12. Brown, 47; Michaelson, Ky, interview with the author, April 2012.
13. Or it may have been $1.4 million. Montandon, 53.
14. Brown, 56.

CHAPTER 14: THE PRETTY BIRD SAGA

1. Brown, 66.
2. Ibid, 67.

3. *Amarillo Globe-News*, "Lawsuit Seeks to Unravel Mystery of Lost Rocket Belt," July 25, 1999. http://amarillo.com/stories/072599/tex_LD0854.001.shtml, accessed October 31, 2012. Hereafter, "Amarillo Globe-News."
4. *Amarillo Globe-News*; Brown, 69. The amount they claimed to have invested in the venture also varied from time to time in different tellings.
5. ABC News, "The Twisted Tale of a Missing Rocket Belt," *Primetime*, October 10, 2002, retrieved on September 1, 2011, www.abcnews.go.com; hereafter "ABC."
6. Brown, 71.
7. Ibid, 74.
8. Ibid, 74–75.
9. *Amarillo Globe-News.*
10. *Amarillo Globe-News*; Brown, 81–82.
11. *Amarillo Globe-News.*
12. Brown, 85.
13. Suitor, interview.
14. Ibid.
15. See David Hambling, "Is It a Bird? Is It a Plane?" *Guardian*, September 13, 2000.
16. *Amarillo Globe-News.*
17. That the flight was no longer than twenty-three seconds is from Suitor. The twenty-eight-second flight was reported in "Rocket Belt Rights Returned," *USA Today*, November 23, 1999. The twenty-eight-second flight time was reported in several other places as well. Hereafter, "*USA Today*." For the longer flight claim, see David Hambling, "Is It a Bird? Is It a Plane?" *Guardian*, September 13, 2000.
18. Brown, 90–91.
19. *Times Daily* (Alabama), "Rocket Belt Designer Wins Lawsuit," July 28, 1999; Pauline Arrillaga, "Where's the Rocket Belt?" Wilmington, NC, *Star-News*, July 27, 1999.
20. Brown, 94–98.
21. *Amarillo Globe-News.*
22. Brown, 118.
23. Barker's excuse for not appearing in court does not seem believable. Even if the notices were all sent to the wrong address, he admitted he knew about the trial taking place within days. There is a short window of time after a trial ends for a party to file an appeal. Barker took no steps to object to the judgment entered against him nor did he file an appeal, or notify the court of any problems with the trial notices.
24. *Times Daily* (Alabama), "Rocket belt designer wins lawsuit," July 28, 1999.
25. ABC.

26. Ibid.
27. Ibid.
28. Ibid.
29. Ibid.
30. The movie *Pretty Bird* was made based upon this rocket belt and its builders. It starred Billy Crudup and Paul Giamatti and was released in 2008.
31. Montandon, 206–7.

CHAPTER 15: CIVILIAN ROCKET BELTS

1. Czaplyski, "Oldies & Oddities."
2. Suitor, *Manual,* 17.
3. Ibid.
4. Dan Schlund, interviewed by the author, May 2012. All references to Schlund are from this interview unless otherwise noted.
5. See www.rocketman.tv/history.php, retrieved April 11, 2012. On the site, it suggests that the company has made "700+" flights but when the link to the "complete list" is clicked, the viewer is directed to a "Contact Us" page.
6. www.rocketman.tv/services.php, retrieved April 30, 2012.
7. www.youtube.com/watch?v=6s-830ScbPo, retrieved April 30, 2012.
8. See www.danschlund.com.
9. Sheri Gibson-Frusher to Shari Sheffield, Letter dated September 1, 2004, hereafter, "Gibson-Frusher to Sheffield." The documents may be found by running a search on the US Patent Office website of the term "Rocketbelt;" See www.uspto.gov.
10. Gibson-Frusher to Sheffield. Typographical errors are in original.
11. Powerhouse Productions, Inc., "Introducing the Rocketman," a tri-fold brochure. The documents may be found by running a search on the US Patent Office website of the term "Rocketbelt." http://tsdr.uspto.gov/documentviewer?caseId=sn78353060&docId=IPC20040913085801, accessed October 31, 2012.
12. David Burge, "The Edison of Crazy," *Garage* 16, 58.
13. Phil Burgess, "Rocket Roundup," retrieved April 22, 2012, www.nhra.com. Hereafter, "Burgess."
14. Burgess.
15. Ky Michaelson, interview with the author, April 2012.
16. "ROCKETBELT," Trademark Registration No. 3,003,697, Dated October 4, 2005; "ROCKETMAN," Trademark Registration No. 3,579,081, Dated February 24, 2009.
17. This writer has seen one such letter that was shown to him in confidence. Recipients of threatening letters from lawyers are often leery of doing anything to stir the pot any further.

18. Correspondence between author and David Joque, "contracts administrator" at Parker in May 2012.
19. Gerard Martowlis, interviewed by the author, May 2012.
20. Martowlis, interview.
21. Ibid.
22. Suitor, *Manual*, 18; Schlund, interview.
23. Michaelson, interview.
24. Thunderbolt Aerosystems "Scientific Team," www.thunderman.net/about/team_suitor.php, retrieved April 9, 2012.
25. Arnold Neracher, correspondence with the author, April 2012. Translated by Jill Warner Dailey.
26. Information from the company's website, www.peroxidepropulsion.com, retrieved April 20, 2012.
27. Montandon, 89.
28. Michaelson, interview.
29. Ibid.
30. Ibid.
31. Suitor, *Manual,* 13.
32. Douglas Brown, "Wanting to Jet into the Future," *Denver Post*, July 17, 2007.
33. BBC, "M25 Chaos after Lorry Explosion," August 30, 2005, retrieved April 20, 2012 from www.news.bbc.co.uk.
34. BBC, "London Rocked by Terror Attacks," July 7, 2005, retrieved April 22, 2012, from www.news.bbc.co.uk.
35. BBC, "7/7 inquests: Coroner Warns over Bomb Ingredient," February 1, 2011, retrieved April 23, 2012, from www.news.bbc.co.uk.
36. BBC, "Six Accused of London Bomb Plot," January 15, 2007, retrieved April 23, 2012 from www.news.bbc.co.uk.
37. Thomas Frank, "TSA: Airline Ban on Liquids Won't Be Lifted Soon," September 9, 2009, retrieved April 28, 2012, www.usatoday.com.
38. BBC, "Three Guilty of Airline Bomb Plot," September 9, 2009, retrieved April 23, 2012, www.news.bbc.co.uk.
39. "Make Your Trip Better Using 3-1-1," retrieved April 23, 2012, www.tsa.gov.
40. Lozano told the author the rocket belts flew twice a day at the expo; Montandon suggests Lozano told him it was four times daily. These kinds of details may not be important although they suggest that some of these mythical origin stories might be fiction mixed with fact. Some sources quote Lozano saying he saw Bill Suitor at this event. If it was in 1962 as Lozano has often said, that would have been two years before Suitor worked for Bell.
41. Troy Widgery's company bought one. TroyWidgery, interview with the author, 2012.

42. Montandon, 160.
43. Ibid, 159.

CHAPTER 16: THE MODERN ROCKET BELT IN THE PUBLIC EYE

1. Peter Gijsberts, "First International Rocketbelt Convention," press release, August 31, 2006.
2. Kevin Purdy, "Airshow: Rocket Men Unite," *The Journal-Register*, August 11, 2007; hereafter, "Purdy."
3. Doug Malewicki, correspondence with the author, September 6, 2011.
4. Spencer, interview.
5. Details of the rocket belt convention are from Peter Gijsberts, interview with the author on numerous occasions, 2011–2012; and Montandon, 85.
6. Purdy.
7. Larry Smith, "Rocket Men," *Slate,* October 4, 2006, retrieved April 27, 2012, www.slate.com.
8. This list can be assembled from various sources but I got mine double-checking with Peter Gijsberts.
9. Larry Smith, interviewed by the author, July 2012,
10. Montandon, correspondence with author, May 2012.
11. Mark Fraunfelder, "Rocketbelt Convention," posted August 7, 2007, retrieved April 20, 2012, www.boingboing.net.
12. Purdy.
13. Gijsberts, correspondence with the author, 2011–2012.
14. Purdy.
15. Ibid.
16. Ibid.
17. The suit named a company apparently formed by Scott, Xtreme Rocket Services LLC, but even that company was never served with the suit; as a result, the case never proceeded against them. See "Memorandum Opinion & Order Granting Defendant's Motion for Summary Judgment," Powerhouse Productions, Inc. et al v. Troy Widgery et al, No. 4:07-cv-071 (E. D. Tex. 2007). Hereafter, "Powerhouse Productions"
18. Powerhouse Productions, 5.
19. It appears that the order denying the appeal was signed by the court on January 26, 2009.
20. "Ditch the Car and Fly to Work with a Jet-Pack," posted December 14, 2007, retrieved April 15, 2012, www.techradar.com
21. Evan Ackerman, "Welcome to the Future, Here's Your Jetpack," June 21, 2007, retrieved April 15, 2012, www.ohgizmo.com.

22. Troy Widgery, interviewed by the author, April–August 2012.
23. Thunderbolt Aerosystems, "Thunderbolt Aerosystems Inc. Unveils ThunderPack," press release, January 24, 2008.
24. Carmelo "Nino" Amarena, correspondence with the author, April–May 2012.
25. Suitor, *Manual,* 85.
26. Bill Houghtaling, interviewed by the author, 2012.
27. Erik Sofge, "The Inside Story of When Jet Packs Really Are Coming," *Popular Mechanics*, October 2009. Hereafter, "Sofge."
28. Sofge.
29. James Lee, "Start," *Wired*, December 2006. Widgery gave a figure of $100 for a five-gallon fuel load when they manufactured it themselves.
30. "Explosion i Aspereds industriomrade," July 23, 2010, retrieved on April 20, 2012, www.gp.se. Translated to the English by Markku Jaakkola and Seija Usitalo, Farmington, Michigan.
31. There was an announcement made on the Peroxide Propulsion website about the accident that was subsequently taken down. Copies of that announcement can be found in places where it was re-posted by hobbyists. One such site is www.hobbyspace.com, retrieved April 20, 2012.
32. Erik Bengtsson, correspondence with the author, May 2012.
33. Sofge.
34. Ibid.
35. Ibid.
36. Spencer, interview.
37. Amarena, correspondence.
38. Widgery, interview.

CHAPTER 17: DUCTED FAN LIFT DEVICES

1. CNN Tech, "Personal 'Jetpack' Gets off the Ground," February 6, 2002, retrieved April 10, 2012 from www.articles.cnn.com.
2. Michael Moshier, "Single Passenger Aircraft," U.S. Patent 6,488,232, filed December 16, 1998 and issued December 3, 2002.
3. Hambling.
4. Ibid.
5. Bulaga, interview.
6. www.trekaero.com, retrieved April 10, 2012.
7. Glenn Neil Martin, "Propulsion Device," US Patent 7,484,687, filed October 26, 2005; issued February 3, 2009. Hereafter, "Martin Patent."
8. Martin Patent, first page ("Abstract").
9. Gregory Mone, "The DIY Flier," *Popular Science*, December 2008, 77. Hereafter, Mone, "DIY Flier."

10. Dan Glaister, "First 'Practical Jetpack' Clears for Take-Off," *Guardian*, July 29, 2008. Hereafter, "Glaister."
11. These notions were raised by several people interviewed by the author, including both rocket belt pilots and aerospace engineers.
12. Mone, "DIY Flier," 78. This writer contacted the author of the *Popular Science* piece to find out if any insight could be gained as to what the record was. The writer indicated that he had simply accepted Martin's word for it and was not aware of anything about the record beyond what he had written. An e-mail directly to the editors of *Popular Science* went unanswered.
13. Bulaga, interview.
14. Mone, "DIY Flier," 119, describes the flight at Oshkosh. There are videos of the flight on www.youtube.com and elsewhere. There, you can clearly see that the device was not free-flown, but was held by two ground personnel through its entire "record-breaking" hover.
15. "Taking Off," The Official Newsletter for the Martin Jetpack, February 2011. Available at www.martinjetpack.com. Hereafter, "Martin newsletter."
16. Martin newsletter.
17. Glaister.
18. Martin newsletter.
19. Alan Boyle, "Is This Your Jetpack?," *Cosmic Log*, retrieved September 10, 2012, http://cosmiclog.nbcnews.com/_news/2008/07/29/4351508-is-this-your-jetpack?lite.
20. Bulaga, interview.
21. Voss, interview.
22. Mone, "DIY Flier," 78.
23. Martin newsletter.
24. Martin newsletter. The $100,000 figure was often given to the press. See Glaister.

CHAPTER 18: HAL GRAHAM

1. Brantley Hargrove, "Hero Pilot Hal Graham's Hard Fall to Earth," *Nashville Scene*, November 26, 2009, www.nashvillescene.com/nashville/hero-pilot-hal-grahams-hard-fall-to-earth/Content?oid=1203877. accessed October 31, 2012. Hereafter "Hargrove."
2. Hargrove.
3. Ibid.

CHAPTER 19: JETLEV

1. All facts regarding Li and the Jetlev are from an interview by the author in 2012 unless otherwise noted.

2. Raymond Li, "Personal propulsion device," US Patent 7,258,301, filed March 23, 2005; issued August 21, 2007. This was filed in 2005 but he had filed his first US patent application in 2004. Hereafter, "Jetlev patent."
3. Jetlev patent, "Background of the invention."
4. Facts and figures on the Jetlev 200 are from the company's website, www.jetlev.com, retrieved April 20, 2012.

CHAPTER 20: JETMAN: YVES ROSSY

1. Kate Ravilious, "'Jet Man' Crosses English Channel Like a Human Rocket," retrieved May 7, 2012, www.nationalgeographic.com. Data on the Jet-Cat engine from the company's website and correspondence with Jet-Cat USA. See www.jetcatusa.com.
2. www.jetman.com/history, retrieved May 6, 2012.
3. "Pictured: Rocketman Flies over Alps with Jet-Pack Strapped to his Back," *Mail Online*, May 15, 2008, retrieved May 6, 2012, www.dailymail.co.uk.
4. Ravilious.
5. Ibid.
6. Ibid.
7. BBC News, "Jetman Yves Rossy Fails in Africa-Europe Flight Attempt," November 25, 2009, http://news.bbc.co.uk/2/hi/europe/8378974.stm, retrieved May 6, 2012.
8. Felica Fonseca, "Swiss Jetman Cancels Grand Canyon Flight," *Deseret News*, May 6, 2011.
9. "JetMan Pulls Off Grand Canyon Flight—Quietly," May 10, 2011, retrieved May 6, 2012, www.msnbc.msn.com.
10. "'Jetman' Soars over Rio de Janeiro: Big Vid," May 7, 2012, retrieved May 7, 2012, www.news.discovery.com.
11. CBS, *Late Show with David Letterman*, May 4, 2012.
12. Yves Rossy, correspondence with author, April, August 2012.

CHAPTER 21: WHEN WILL WE HAVE JET PACKS?

1. Purdy.

EPILOGUE: WHERE ARE THEY TODAY?

1. DuBarry, 104.
2. "The De Lackner Aerocycle—An Early Flying Platform," www.transchool.lee.army.mil, retrieved April 4, 2012.
3. www.nasm.si.edu. Search for "rocket belt." Retrieved January 23, 2012. In their description, they refer to it as a type of "Jet Pack."

4. Check at www.wnyaerospace.org. As of this writing, May 8, 2012, the museum is closed and no one has returned e-mails sent by the author, seeking further information from the museum.
5. Lindenbaum, 2:59.
6. Montandon, 206–7.
7. Lindenbaum wrote that two were built. Lindenbaum, 3:42. Two were required for the contract, but Williams built three.

APPENDIX: ORIGIN HOAXES

1. www.gizmodo.com/5524604/real-bloody-flying-nazi-soldiers-with-jet-packs.
2. The author is aware that there were, in fact, piloted V-1 buzz bombs flown during World War II. Even so, the notion of wearing a pulse-jet engine is very different from the notion of flying an aircraft powered by one.
3. Brown, 16.
4. Correspondence by author with operators of various websites. One admitted the idea was a "fantasy." Another indicated that he had "heard" of these devices since the 1970s but had no evidence to support their existence. Even so, he believed they were real.
5. The author encountered this argument more than once while trying to run down the facts on this. That is, the lack of evidence to support the idea is proof of its existence.
6. Jerome Clark, *Unexplained: Strange Sightings, Incredible Occurrences & Puzzling Physical Phenomena* (Farmington Hills, MI: Visible Ink Press, 1999), 558.
7. Frank Miele, "Giving the Devil His Due," *Skeptic*, Vol. 2, No. 4, 60.
8. "Justin Capra—National Geographic Channel," uploaded May 30, 2008, https://www.youtube.com/watch?v=OgbDrXiOPK4.
9. "Părintele primului rucsac zburător," www.adevarul.ro/actualitate/Parintele-primului-rucsac-zburator_0_35998137.html.
10. See www.youtube.com/watch?v=jz_4k51eeGw. The title of the video translates roughly as "50th anniversary of the first flight made by individual flight device – Justin Capra."
11. www.youtube.com/watch?v=KR8R9-2Emng.
12. "Justin Capră: inventator, măturător, filozof şi pustnic," www.gandul.info/reportaj/justin-capra-inventator-maturator-filozof-si-pustnic-2380119.
13. Correspondence with Calin Dinulescu.
14. The piece's title in English is "Remember Those Flying." Acei zburători uitaţi, www.natgeo.ro/istorie/personalitati-si-evenimente/9257-acei-zburatori-uitati.
15. Cecil Martin and Robert L. Cummings, "Turbo-Fan Lift Device," US Patent 3,023,980, filed October 13, 1958 and issued March 6, 1962.

16. www.gandul.info/reportaj/justin-capra-inventator-maturator-filozof-si-pustnic-2380119.
17. *Romanian Panorama*, 1995, 23.
18. "Interview with Justin Capra," www.imaginisiganduri.blogspot.com/2008/02/interviu-cu-justin-capra.html.
19. See www.youtube.com/watch?v=jz_4k51eeGw, at the 2:16 mark. The title of the video translates roughly as "50th anniversary of the first flight made by individual flight device – Justin Capra."
20. Radio Romania International's website published a story on Capra and likewise stated, without reservation, that his was "the first jet-pack." www.rri.ro/art.shtml?lang=1&sec=170&art=262142.
21. One such website is www.pancuantic.ro/fundatie/justin.html. The site is written in Romanian but translations into English describe Capra's "'Flying backpack' invention that was stolen and patented seven years later by the Americans." Another website quotes Capra stating that the plans for the jet pack may have been stolen by "spies" or Romanian "agents," and sold to the Americans. www.financiarul.ro/2011/05/18/inventatorul-roman-justin-capra-a-creat-masina-care-consuma-05-litri-la-suta-de-kilometri-poti-sa-faci-toata-tara-cu-ea-cu-un-singur-plin/.
22. www.arec.ca/-famouspeople/.

INDEX